Springer Tracts in Modern Physics
Volume 175

Springer-Verlag Berlin Heidelberg GmbH

Springer Tracts in Modern Physics

Springer Tracts in Modern Physics provides comprehensive and critical reviews of topics of current interest in physics. The following fields are emphasized: elementary particle physics, solid-state physics, complex systems, and fundamental astrophysics.
Suitable reviews of other fields can also be accepted. The editors encourage prospective authors to correspond with them in advance of submitting an article. For reviews of topics belonging to the above mentioned fields, they should address the responsible editor, otherwise the managing editor.
See also http://www.springer.de/phys/books/stmp.html

Hiroshi Kudo

Ion-Induced Electron Emission from Crystalline Solids

With 85 Figures

Springer

Dr. Hiroshi Kudo
University of Tsukuba
Institute of Applied Physics
305-0006 Tsukuba, Ibaraki, Japan
E-mail: kudo@bukko.bk.tsukuba.ac.jp

Library of Congress Cataloging-in-Publication Data.
Die Deutsche Bibliothek - CIP-Einheitsaufnahme
Kudo, Hiroshi:
Ion-induced electron emission from crystalline solids/Hiroshi Kudo.

(Springer tracts in modern physics; Vol. 175)
(Physics and astronomy online library)
ISBN 978-3-662-14661-3 ISBN 978-3-540-45527-1 (eBook)
DOI 10.1007/978-3-540-45527-1

Physics and Astronomy Classification Scheme (PACS): 79.20.Rf, 61.85.+p, 34.50.Dy

ISSN print edition: 0081-3869
ISSN electronic edition: 1615-0430
ISBN 978-3-662-14661-3

http://www.springer.de

Originally published by Springer-Verlag Berlin Heidelberg New York in 2002

Typesetting: Camera-ready copy from the author using a Springer LaTeX macro package
Cover design: *design & production* GmbH, Heidelberg

Printed on acid-free paper SPIN: 10842543 57/3141/tr 5 4 3 2 1 0

Preface

This monograph deals with ion-induced electron emission from crystalline solids bombarded by fast ions. During the past decade, electron spectroscopy combined with the ion channeling technique has revealed various "messages" about ion–solid and electron–solid interactions carried by the emitted electrons. While the ion-induced electrons produced by binary-encounter processes are of primary interest in this book, closely related topics such as the emission of ion-induced Auger electrons from crystal targets are also reviewed, with emphasis on their interdisciplinary aspects, for example, their relation to photoelectron diffraction. In addition to these topics, the book describes the underlying physics and experimental techniques so that it should provide useful information for students and scientists working in ion-beam-based research and development in various areas of atomic and solid-state physics, materials science, surface science, etc.

I am much indebted to the gererations of students who have passed through my laboratory, since they have stimulated me with elementary but essential questions in various phases of the studies. I am also grateful to T. Azuma, Y. Kido, K. Kimura, H. Naramoto, and S. Seki for critical reading of the manuscript.

Tsukuba, August 2001 *Hiroshi Kudo*

Contents

1. Introduction

Ion-induced electron emission from solids reflects a variety of physical processes such as electronic excitation and ionization in the target material, the transport of energetic electrons through solids, electron capture and loss by the projectile during its passage, etc. For over a hundred years many workers have studied this phenomenon not only out of pure physical interest but also for practical purposes, for example various applications to surface layer analysis of materials. These numerous experimental and theoretical studies have been reviewed in a number of articles [1–5]. Also, the atomic processes responsible for ion-induced electron emission have been reviewed extensively [6,7].

While the solid targets used in most of these studies were polycrystalline, observations using single-crystal targets have been reported since the early 1960s [8–12]. In these studies, reduced electron yields were observed when the ions were incident in transparent, i.e. channeling, directions of the crystal lattice. It is notable that this period overlaps the early days of ion channeling, which later became a key concept in charged-particle interactions with single crystals [13–15]. Furthermore, some groups reported observations of diffraction effects of ion-induced low-energy electrons emitted from a crystal target [16–19]. Now, the diffraction and related effects need to be discussed in a unified manner, for example in connection with photoelectron diffraction, which has been widely used for surface structure analysis.

The next phase of the study of ion-induced electron emission from crystal targets started in about 1980 when several groups measured ion-induced Auger electrons under channeling incidence conditions [20–22]. The interest of these groups included the applicability of the channeling technique to ion-induced Auger electron spectroscopy for surface structure analysis. The observed Auger energy spectra typically exhibited a reduced peak height and a short low-energy tail under channeling incidence conditions. These spectra were successfully accounted for in terms of the ion beam shadowing effect. However, the reduced continuum electron yields generally observed in the experiments were explained less quantitatively [20] or attracted only minor attention [22].

The subsequent progress in the experimental studies of the continuum electron yield from crystal targets is the primary topic of this monograph. These studies rely entirely on the well-established channeling technique, which enables control of impact parameters in the ion–atom collision. Measurements of the continuum electron yield under channeling as well as nonchanneling incidence conditions have been carried out since the late 1980s, mainly at the University of Tsukuba and at the Japan Atomic Energy Research Institute at Takasaki, using ion beams obtained from accelerators of the tandem and single-stage types, and from an ion implanter as well. The experimental data cover the ion energy range 0.1–10 MeV/u, ion species of atomic numbers from 1 to 17, and various charge states. The electron measurements at $\sim$180° with respect to the ion beam direction allowed practical data acquisition with solid targets. The crystal targets used include metals, semiconductors, and insulators.

The present understanding of ion-induced electron emission from crystal targets can be outlined as follows:

- Ion-induced electron emission reflects the initial stage of ion channeling, i.e. gentle deflection of the incident ions away from the aligned atoms in the crystal, which causes reduced electron emission in the backward direction. This is clearly seen for the electrons in the keV energy range, which were originally produced by binary-encounter processes.
- The electron emission results not only from close ion–atom collisions but also from distant collisions. Accordingly, the electron emission under channeling incidence conditions is sensitive to the ion's behavior relatively far from the aligned atoms, in contrast to close-encounter phenomena such as ion backscattering or inner-shell X-ray emission.
- The electron yield in the channeling case can be accounted for in terms of the effective thickness of the unshadowed surface layer and the effective escape length which characterizes the electron yield in the nonchanneling case. These quantities can be estimated from experiments, and partly from calculations. Detailed knowledge of the escape processes of the emitted electrons, including further production of energetic secondary electrons, is unnecessary in most of the applied studies.

Obviously, the underlying physics includes ionization by binary collisions, ion channeling, and elastic as well as inelastic scattering of energetic electrons in solids. The first two topics are reviewed in Chaps. 3 and 4, respectively. In particular, Chap. 4 introduces ion channeling in an unfamiliar manner, i.e. starting from the Coulomb shadow, not from the continuum potential. This is because we focus mainly on the effective number of target atoms that are shadowed and contribute very little to the ion-induced electron emission. At present, most of the atomic collision phenomena under any given experimental condition can be well reproduced by numerical calculations. Therefore, the main role of the analytical models presented in Chaps. 3 and 4 is to provide the general and essential aspects of the underlying physics, even though they are rather qualitative.

Chapter 5 reviews the experimental apparatus and techniques for precise and reliable measurements of the binary-encounter and Auger electron yields from a crystal target. This is intended to allow well-focused discussion in the subsequent chapters within the restricted boundaries imposed by the measurement conditions. The backward electron emission from solid targets is discussed in Chap. 6. Chapter 6 is intended to provide knowledge for understanding the simpler aspects of binary-encounter electron emission from solids, which appears somewhat complex since it results from multi-stage processes including ion–atom, ion–electron, and electron–atom interactions. Chapter 7 reviews studies of monoenergetic electron emission from lattice sites of the crystal, in relation to inner-shell Auger processes. These studies have provided knowledge about the physical processes associated with crystal targets, which provides easier access to an understanding of the less simple processes of continuum electron emission, discussed in Chap. 8. Chapters 9 and 10 are devoted to applications of binary-encounter electron spectroscopy. Chapter 11 briefly describes some related topics, e.g. the emission of low-energy electrons in ion–surface interactions.

Evidently, the present understanding of the phenomena is more comprehensive and precise than in earlier stages of the research. Hence several earlier papers published by the author's group will be reviewed, with appropriate refinement based on the present better knowledge about the phenomena.

2. Terminology and Table of Symbols

In this monograph, several technical terms and physical concepts are used that do not necessarily follow the common usage in the fields of electron spectroscopy and ion backscattering spectroscopy. Rather, they are used in a modified or extended manner to describe specific topics. The following notes and the table will be useful to avoid any confusion that may arise from terminology, especially for those who are already familiar with either electron spectroscopy or ion channeling.

2.1 Notes on Terminology

1. The term *high-energy shadowing effect* is preferentially used throughout this monograph rather than the more popular term *channeling effect*. This is due to the fact that we are more interested in the shadowed and unshadowed target atoms than in the behavior of channeled ions.
2. Another reason for use of "high-energy shadowing effect" is to avoid confusion with the familiar shadowing effect for low-energy ions. In the former the effective shadow results from ion–atom multiple scattering, while in the latter it results from single scattering.
3. In this book the term *electron yield* is used for the electron emission intensity from a solid target as a function of the detection angle and emission energy. It is not used, as is often the case, for the total electron yield per incident projectile emitted from a solid surface, unless otherwise noted.
4. The term *random* is frequently used to mean *nonchanneling*, following the customary usage in the channeling and related fields.
5. Also following the customary usage, $\langle hkl \rangle$ and (hkl) are used to indicate a channeling axis and plane, respectively. The notation $[hkl]$ is also used when channeling axes must be specified. In other cases, the notations $\langle hkl \rangle$, $[hkl]$, $\{hkl\}$, and (hkl) follow the common usage in crystallography.
6. In most cases, the ion energies are represented in MeV/u, since the electron emission induced by fast ions is essentially velocity dependent.

2.2 Frequently Used Symbols

a	Screening length of an atomic potential: (4.29), (4.30), (4.31)
D	Effective escape length: Sect. 6.2
d	Interatomic spacing along a crystal axis
E_{B}	Binary-encounter peak energy: (3.2)
E_{L}	Loss-peak energy: (6.2)
e	Electronic charge, $e^2 = 14.4\ \mathrm{eV\,Å}$
m_{e}	Rest mass of electron
R_{c}	Shadow cone radius at $z = d$: (4.26)
t	Effective surface layer thickness for production of binary-encounter electrons: Sect. 8.4.1
V_1, E_1	Ion velocity and kinetic energy in the laboratory frame
W	Normalized channeling yield, i.e. the ratio of the channeling to the random (nonchanneling) electron yield: Sect. 8.2
Z_1, M_1	Atomic number and mass of projectile ion
Z_2, M_2	Atomic number and mass of target atom
Z_{eff}	Effective nuclear charge for high-energy shadowing: Sect. 9.2
$\Delta E/E$	Relative energy resolution of spectrometer: (5.2)
θ, ϕ	Tilt angles of the crystal
λ	Electron mean free path for inelastic scattering: Sect. 7.2.2
λ_{t}	Electron transport mean free path: Sect. 7.2.3
μ	Electron yield originating from unshadowed valence and loosely bound electrons: (8.1)

3. Binary-Encounter Electron Emission

Bombardment of atoms by fast ions gives rise to the emission of energetic electrons by kinetic processes. These electrons are distributed over a wide energy range, and hence provide a continuum energy spectrum which generally has no sharply defined high-energy end. High-energy electrons are emitted preferentially in the forward direction, as is clearly observed in experiments using gas targets. The emission of high-energy electrons is of primary interest here since it is sensitive to the high-energy shadowing effect in a crystal target. The hard collision processes producing high-energy electrons can be described by the binary-encounter model, which was developed essentially from the early work described in "Ionization by moving electrified particles" by Thomson [23]. This chapter presents an outline of the binary-encounter theory and some notable results useful for the discussion in later chapters.

3.1 Ion–Electron Elastic Collisions

We first discuss an ion–electron elastic collision in a laboratory frame where a free electron of mass m_e initially at rest is recoiled by a nonrelativistic ion. For simplicity, the projectile is assumed to be a fully stripped ion of velocity V_1, atomic number Z_1, mass M_1, and kinetic energy $E_1 = M_1 V_1^2/2$. The conservation of energy and momentum leads to an expression for the energy T transferred from the ion to the electron. This energy is given as a function of the recoil angle ϕ_1 with respect to the incident direction of the ion [7, 24]:

$$T = \frac{4 m_e M_1 E_1}{(m_e + M_1)^2} \cos^2 \phi_1 \simeq 2 m_e V_1^2 \cos^2 \phi_1 \,, \tag{3.1}$$

under the condition $m_e \ll M_1$; ϕ_1 is in the range $0 \leq \phi_1 \leq \pi/2$. The maximum transferred energy E_B, corresponding to a head-on collision ($\phi_1 = 0$), is therefore given by

$$E_B = 2 m_e V_1^2 \,. \tag{3.2}$$

E_B is the *binary-encounter peak energy* [25], named after the pronounced peak typically observed at this energy value in the electron energy spectrum measured in the forward direction with a gas target (Sect. 3.6.1).

Also under the condition $m_e \ll M_1$, the Rutherford differential cross section as a function of ϕ_1 is given by

$$\frac{d\sigma_R}{d\phi_1} = 2\pi \left(\frac{Z_1 e^2}{m_e V_1^2} \right)^2 \sec^3 \phi_1 \sin \phi_1 \ . \tag{3.3}$$

By changing the parameter from ϕ_1 to T, using (3.1), the energy distribution of the recoiled electrons $\mathcal{Y}_0$ is obtained:

$$\mathcal{Y}_0(T) = \frac{d\sigma_R}{dT} = \frac{4\pi Z_1^2 e^4}{E_B} \frac{1}{T^2} \ . \tag{3.4}$$

T lies in the range $0 \leq T \leq E_B$, which corresponds to a range of ϕ_1 from $\pi/2$ down to 0, according to (3.1). For ions in the MeV/u energy range, most of the valence and loosely bound atomic electrons in solids can be regarded as particles at rest, and therefore the energy transfer to these electrons can be approximately described by (3.4).

For atomic electrons that have high orbital velocities, however, the maximum transferred energy can be greater than E_B. This is due to the possible increase in the relative velocity. In the binary-encounter model, the momentum transfer to an atomic electron results from the interaction of the ion with a free electron moving at the same speed as the orbital velocity of the electron. The effect of binding is taken into account after the momentum transfer.

Two special cases are considered here. When an electron with an orbital velocity v_2 is moving antiparallel to the ion velocity, the maximum transferred energy in a head-on collision, T_+, is given by

$$\begin{aligned} T_+ &= m_e(2V_1 + v_2)^2/2 - m_e v_2^2/2 \\ &= E_B + 2m_e V_1 v_2 \ . \end{aligned} \tag{3.5}$$

We further define another parameter to be used in Sect. 3.2, the transferred energy T_- which corresponds to the opposite case, in which the initial electron velocity is parallel to the ion velocity:

$$T_- = E_B - 2m_e V_1 v_2 \ . \tag{3.6}$$

The expression for T_- is valid under the condition $V_1 > v_2$, i.e. $T_- > 0$, the condition for the collision to occur. Since the electron must overcome the binding energy $\mathcal{I}$ when leaving the atom, the observed maximum energy E_{max} should be written as

$$E_{max} = T_+ - \mathcal{I} \ . \tag{3.7}$$

As can be seen from the above cases for recoil at 0°, there is no one-to-one correspondence between the tranferred energy and the recoil angle of the orbital electron, unlike (3.1) for electrons at rest. Electron emission from orbitals is therefore described by a double differential cross section with respect to T and ϕ_1. The integral of the double differential cross section with respect to ϕ_1, i.e. over the solid angle, leads to the energy spectrum of the electrons emitted in all directions.

It is worth pointing out that an orbital electron can be recoiled in a backward direction ($\pi/2 < \phi_1 \leq \pi$), in contrast to an electron at rest. This effect is expected to be typical for the recoil of high-velocity orbital electrons caused by a relatively slow ion [7]. In such a case, the orbital electron is ionized with a significant change of the momentum, but with only a small transfer of kinetic energy.

3.2 Recoil Cross Section of Orbital Electrons

We need to obtain an expression corresponding to (3.4) when the electron is not at rest, but has an orbital velocity. To do this, the binary-encounter model assumes Rutherford scattering of the two moving particles, i.e. an incident ion of velocity $\boldsymbol{V}_1$ and an electron of initial velocity $\boldsymbol{v}_2$ in the laboratory frame, as shown in Fig. 3.1. This is essentially an impulse approximation.

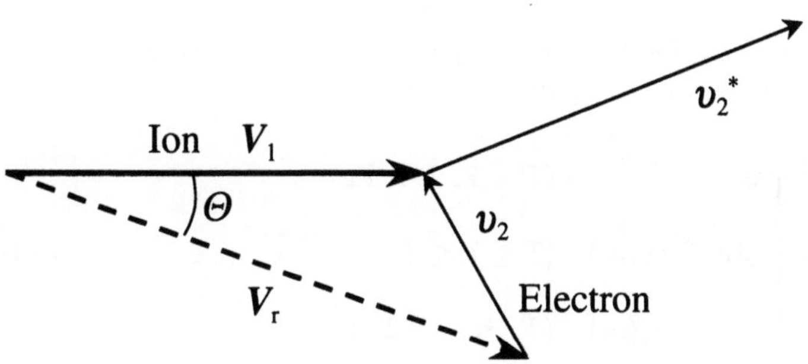

Fig. 3.1. Encounter of two moving particles in the laboratory frame

Under the assumption that the momentum of the ion remains unchanged during the collision in the laboratory frame, the center-of-mass velocity $\boldsymbol{V}_\mathrm{c}$ and the relative velocity $\boldsymbol{V}_\mathrm{r}$ are given by

$$\boldsymbol{V}_\mathrm{c} = (M_1 \boldsymbol{V}_1 + m_\mathrm{e} \boldsymbol{v}_2)/(M_1 + m_\mathrm{e}) \simeq \boldsymbol{V}_1 \ , \tag{3.8}$$

$$\boldsymbol{V}_\mathrm{r} = \boldsymbol{V}_1 - \boldsymbol{v}_2 \ . \tag{3.9}$$

After averaging over the azimuthal angle of the velocity vector of the scattered electron $\boldsymbol{v}_2^*$ with respect to the direction of $\boldsymbol{V}_\mathrm{c}$, i.e. $\boldsymbol{V}_1$ in this case, the energy distribution of the recoiled electron is written as [26]

$$\frac{\mathrm{d}\sigma_\mathrm{R}(\boldsymbol{V}_1, \boldsymbol{v}_2)}{\mathrm{d}T} = \frac{2\pi Z_1^2 e^4 V_1^2}{V_\mathrm{r}^2 T^3} \left(1 - \cos^2 \Theta + \frac{T}{m_\mathrm{e} V_1 V_\mathrm{r}} \cos \Theta \right) , \tag{3.10}$$

under the condition, which is required from collision kinematics,

$$-1 \leq \cos\Theta - T/m_e V_1 V_r \leq 1\,, \tag{3.11}$$

where Θ is the angle between $\boldsymbol{V}_r$ and $\boldsymbol{V}_1$, as shown in Fig. 3.1.[1] The restriction (3.11) guarantees positive values of (3.10). $\mathrm{d}\sigma_R(\boldsymbol{V}_1, \boldsymbol{v}_2)/\mathrm{d}T = 0$ unless (3.11) is satisfied. For $\boldsymbol{v}_2 = 0$, (3.10) coincides with (3.4).

Further calculations include averaging over the direction of $\boldsymbol{v}_2$ assuming an isotropic $\boldsymbol{v}_2$ distribution in the laboratory frame. Furthermore, the calculated result must be expressed in the laboratory frame. Indeed, the main part of the calculation is the presentation of the cross section in terms of the velocities and kinetic energies in the laboratory frame [26–31]. The energy spectrum of the electrons emitted in all directions can be obtained in this manner.

Because of the binding energy of the electron, the observed energy of the electron released from the atom will not be equal to T. It is assumed that when leaving the atom the recoiled electron loses an amount of kinetic energy equal to $\mathcal{I}$, as noted in Sect. 3.1. T is therefore related to the observed electron energy $E_e(\geq 0)$ by

$$T = E_e + \mathcal{I}\,. \tag{3.12}$$

The distribution of E_e, i.e. the energy spectrum of the electrons emitted in all directions $\mathcal{Y}_1(E_e)$, is given by [6,26,27,30]

$$\mathcal{Y}_1(E_e) = \begin{cases} 0 & (T < \mathcal{I},\, T > T_+) \\ \mathcal{Y}_0(T)\, f_A(v_2) & (\mathcal{I} \leq T \leq T_-) \\ \mathcal{Y}_0(T)\, f_B(v_2) & (T_- \leq T \leq T_+) \end{cases}\,, \tag{3.13}$$

where T_+ and T_- have already been defined in (3.5) and (3.6), respectively, and

$$f_A(v_2) = 1 + \frac{2m_e v_2^2}{3T}\,, \tag{3.14}$$

$$f_B(v_2) = \frac{m_e v_2^2}{12T}\left[\left(\frac{2V_1}{v_2}\right)^3 + \left(1 - \sqrt{1 + \frac{2T}{m_e v_2^2}}\right)^3\right]. \tag{3.15}$$

Clearly $\mathcal{Y}_1(E_e)$ coincides with $\mathcal{Y}_0(T)$ when $\mathcal{I}$ and, correspondingly, v_2 decrease to 0.

A refined expression for the electron energy spectrum can be obtained by averaging $\mathcal{Y}_1(E_e)$ over the orbital velocity distribution of the electron $\mathcal{F}(v_2)$.

[1] Equation (3.10) corresponds to a special case, i.e. $m_e \ll M_1$, of Gerjuoy's expression [26], which is applicable to any pair of projectile and target masses.

After careful consideration of the range of the parameter v_2 in the integral that arises from averaging, the refined electron energy spectrum $\mathcal{Y}_2(E_e)$ can be written

$$\mathcal{Y}_2(E_e) = \begin{cases} \mathcal{Y}_0(T)\left[\int_0^{v_c} f_A(v_2)\mathcal{F}(v_2)\,dv_2 + \int_{v_c}^{\infty} f_B(v_2)\mathcal{F}(v_2)\,dv_2\right] \\ \qquad\qquad\qquad\qquad\qquad\qquad (\mathcal{I} < T \le E_B) \\ \mathcal{Y}_0(T)\int_{v_c}^{\infty} f_B(v_2)\mathcal{F}(v_2)\,dv_2 \qquad\qquad (T \ge E_B) \end{cases}, \tag{3.16}$$

where $v_c = |(1 - T/E_B)V_1|$.

3.3 Recoil from a Hydrogen-Like Atom

A further analytical approach is presented for a hydrogen-like atom, which demonstrates essential aspects of the recoil cross section of orbital electrons. We first consider the distribution of the orbital velocity. In the quantum-mechanical treatment, the distribution is identical with the momentum representation of the electron wave function. Fock has thus derived the quantum-mechanical distribution of the orbital velocity for a hydrogen-like atom $\mathcal{F}_0(v_2)$ [32]. The same expression can be also derived classically by assuming the microcanonical statistical distribution of the electron [30], i.e.

$$\mathcal{F}_0(v_2)\,dv_2 = C_0\left(\int \delta\left(\mathcal{I} + m_e v_2^2/2 - Z_2 e^2/r\right) d^3r\right) 4\pi v_2^2\,dv_2\ , \tag{3.17}$$

where $\delta(X)$ represents the delta function for a variable X, and the normalization constant C_0 is determined so that

$$\int_0^{\infty} \mathcal{F}_0(v_2)\,dv_2 = 1\ . \tag{3.18}$$

Equation (3.17) leads to the Fock distribution:

$$\mathcal{F}_0(v_2)\,dv_2 = \frac{32u^5 v_2^2}{\pi(v_2^2 + u^2)^4}\,dv_2\ , \tag{3.19}$$

where u is defined by

$$m_e u^2/2 = \mathcal{I}\ . \tag{3.20}$$

$\mathcal{F}_0(v_2)$ is shown as a function of $q = v_2/u$ in Fig. 3.2.

It should be noted that the quantum-mechanical expectation value of the electron kinetic energy is given by

$$\frac{m_e}{2}\int_0^{\infty} v_2^2 \mathcal{F}_0(v_2)\,dv_2 = \frac{m_e}{2}u^2 = \mathcal{I}\ , \tag{3.21}$$

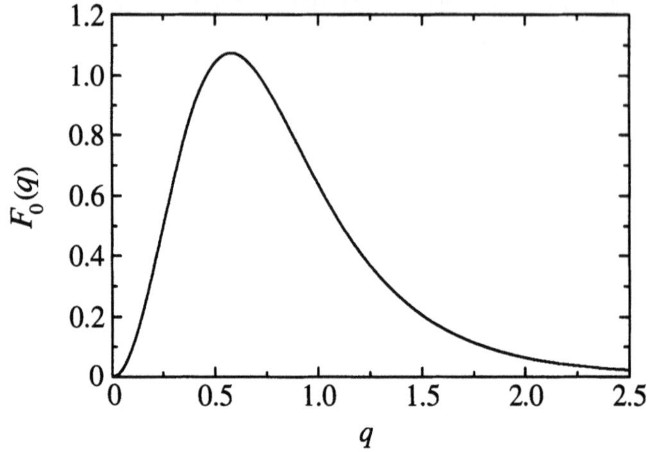

Fig. 3.2. Fock distribution (3.19) shown as a function of $q = v_2/u$

indicating a specific relation between the mean kinetic energy and the binding energy for a hydrogen-like atom. The integral of $\mathcal{F}_0(v_2)$ for $v_2 \geq u$ is equal to 0.29, implying that there is a certain probability for the maximum transferred energy to be greater than E_B, as mentioned in Sect. 3.1. Also, there exists no upper limit of the transferred energy, because of the smooth tail towards the high-velocity side in the Fock distribution.

$\mathcal{Y}_2(E_\mathrm{e})$ for a hydrogen-like atom can be calculated from (3.16) if we take $\mathcal{F}(v_2) = \mathcal{F}_0(v_2)$. The integrals involved in (3.16) can be performed analytically and expressed with the two dimensionless parameters α and ω, defined by

$$\alpha = T_\mathrm{e}/\mathcal{I} = 1 + E_\mathrm{e}/\mathcal{I}\,, \qquad \omega = E_\mathrm{B}/\mathcal{I}\,.$$

For presentation of the results obtained by Rudd et al. [33], we use, further, the following notations:

$$\beta = \omega\left(1 - \frac{\alpha}{4\omega}\right)^2, \quad \alpha_1 = \alpha - 1\,, \quad \beta_1 = 1 + \beta\,, \quad \tau = \alpha + \beta\,.$$

Note that these six parameters always take positive values. The final result can be written

$$\mathcal{Y}_2(E_\mathrm{e}) = \begin{cases} \mathcal{Y}_0(T)\,(\mathcal{S}_1 + \mathcal{S}_2)/4\pi\alpha & (\mathcal{I} < T \leq E_\mathrm{B}) \\ \mathcal{Y}_0(T)\,\mathcal{S}_2/4\pi\alpha & (T \geq E_\mathrm{B}) \end{cases}, \tag{3.22}$$

where

$$\mathcal{S}_1 = \frac{32\beta^{3/2}\alpha}{3\beta_1^3} + \left(\frac{4}{3} + \alpha\right)(\pi - 2\mathcal{R}_1) ,$$

$$\mathcal{S}_2 = \frac{16}{3\beta_1^3}\left(\frac{4}{3}\omega^{3/2} - \beta^{3/2}\alpha + \frac{\alpha\tau^{3/2}}{\alpha_1}\right) + \left(\frac{4}{3} + \alpha\right)\mathcal{R}_1 - \left(\frac{4}{3} + \frac{\alpha}{\alpha_1}\right)\mathcal{R}_2 ,$$

with $\mathcal{R}_1$ and $\mathcal{R}_2$ given by

$$\mathcal{R}_1 = \arctan\beta^{-1/2} + \frac{\beta^{1/2}}{\beta_1^3}\left(1 + \frac{8\beta}{3} - \beta^2\right),$$

$$\mathcal{R}_2 = \alpha_1^{-3/2}\ln\left(\frac{\tau^{1/2} - \alpha_1^{1/2}}{\beta_1^{1/2}}\right) + \frac{\tau^{1/2}}{\beta_1^3}\left(2 + \frac{14\beta}{3} + \frac{8\alpha}{3}\right) + \frac{\tau^{1/2}}{\beta_1\alpha_1} .$$

A comment can be made on the spectrum shape. $\mathcal{Y}_2(E_e)$ generally decreases with increasing E_e, while the double differential cross section, mentioned in Sect. 3.1, has a pronounced binary-encounter peak at an energy approximately given by (3.1), for a forward recoil angle (see also Sect. 3.6.1). The binary-encounter maximum which might be anticipated in the spectrum fails to remain after integration over the electron emission angle. This is due to the larger contribution from the recoiled electrons with lower kinetic energies. A lower-energy electron corresponds to larger ϕ_1 in (3.1). The binary-encounter peak eventually disappears after shifting towards the lower-energy side.

3.4 Z_1^2 Scaling of the Electron Yield

According to (3.16), $\mathcal{Y}_2(E_e)$ can generally be expressed as Z_1^2 multiplied by a function which depends on the parameters $\mathcal{I}$ and E_B. This indicates that, for equal-velocity ions (i.e. for the same value of E_B), $\mathcal{Y}_2(E_e)$ can be scaled with Z_1^2 under the condition that the ions are fully stripped.

Such a Z_1^2 scaling character is widely anticipated for binary-encounter electron emission. This can be concluded simply from the mathematical processes in the transformations of the coordinate system when (3.13) is derived. The physical parameters which are concerned with the transformations between the center-of-mass and laboratory frames include the masses, velocities, and directions of motion the ion and the electron, but never include Z_1. It follows that a Z_1^2 dependence and therefore a Z_1^2 scaling are also anticipated for the double differential cross section in the laboratory frame.

Obviously, the Z_1^2 scaling stems from the pure Coulomb field of the projectile. In the quantum-mechanical perturbation theory of electron emission, the Z_1^2 scaling arises from (the square of) the matrix element of the pure Coulomb interaction. Actually, the Z_1^2 scaling for fully stripped ions is a typical conclusion of general treatments of the differential cross section by the first-order Born approximation, including the stopping power for high-energy ions (the Bethe stopping formula) [24, 34, 35]. A higher-order perturbation should give rise to a correction term of higher-order Z_1 than the Z_1^2 term.

The Z_1^2 scaling does not hold for partially stripped ions whose nuclear charges are screened by bound electron(s). In this case, the recoil cross section is usually modified in a complicated manner, as is well known in zero-degree electron spectroscopy (Sect. 3.6.1). The binary-encounter electron emission from crystal targets bombarded by partially stripped ions will be discussed in Chap. 9.

3.5 Remarks on the Binary-Encounter Model

We recall that the Rutherford cross section can be obtained from a rigorous solution of the Schrödinger equation [36, 37] (see also Sect. 4.2). Therefore, a binary-encounter theory based on the Rutherford cross section inherently includes all the higher-order effects which are neglected in the perturbation theory. In this sense the binary-encounter theory has a general basis, unlike the Born approximation, when strong projectile–electron interactions, as in the present case, are considered [7]. Accordingly the binary-encounter theory provides a crude but general description of the ion-induced emission of atomic electrons. Therefore, calculations by the binary-encounter theory, especially using (3.22), are of practical importance for estimating the production cross section of high-energy electrons (of the order E_B) recoiled from specific inner shells. This allows detailed analysis of the experimental data of the high-energy shadowing effect when the ions are fully or highly stripped in the crystal.

It is important to note that the binary-encounter theory must be applied carefully to partially stripped ions whose nuclei are effectively screeened by the bound electrons. In this case, the phases of the electron scattering amplitude play a significant role in the emission intensity in the forward direction [38]. Such a case will be discussed in Chap. 9.

There are more refined calculations of the electron production cross section, such as the classical-trajectory Monte Carlo method or those based on quantum-mechanical perturbation theory. These are useful for obtaining more precise estimates of the cross sections but in many cases numerical treatments are required [7].

3.6 Experimental Approaches

Numerous experiments using gas targets have been carried out to study the production of ion-induced electrons by binary-encounter processes. Some typical experimental approaches are only briefly described here, for a better understanding of the electron emission phenomena.

3.6.1 Zero-Degree Electron Spectroscopy

Binary-encounter electron emission can be observed in an elegant way in the 0° direction ($\phi_1 = 0$) using gas targets [7]. In zero-degree electron spectroscopy, a pronounced binary-encounter peak appears at an energy equal to $\sim E_B$ in the energy spectrum. A typical example is shown in Fig. 3.3, where the binary-encounter peak for a He gas target bombarded by 56 MeV Si^{6+} is observed near $E_B = 4.36$ keV [39]. The electrons emitted at 0° result from

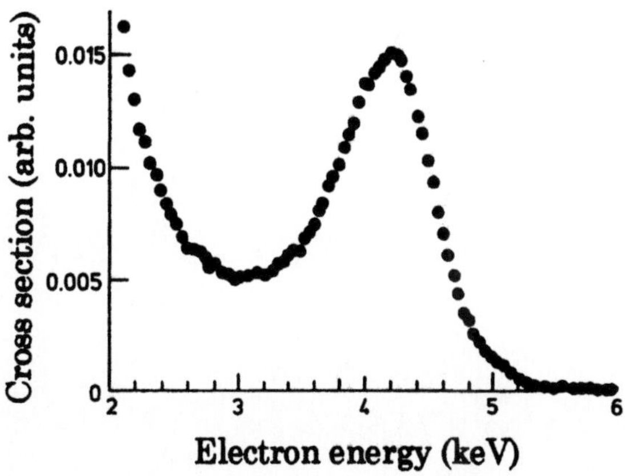

Fig. 3.3. Electron energy spectrum for collision of 56 MeV Si^{6+} + He observed at 0° [39]

head-on collisions which reflect strong two-body interactions between the ion and electron. It is obvious from Sect. 3.1 that the energy spectrum of the emitted electrons at $\phi_1 = 0$ reflects the distribution of the electron momentum parallel to the beam direction before the collision, which is actually the Compton profile of the atomic electrons [40].

Zero-degree electron spectroscopy is equivalent to electron backscattering experiments at 180° in the projectile frame. This observation technique allows one to study, for example, the dependence of the electron backscattering yield at 180° on the charge state of the target ion [7].

3.6.2 Energy Distribution of All Electrons

Studies of the energy spectra of electrons emitted from gas targets bombarded by keV to MeV protons have been summarized in the reviews by Rudd and

Macek [41] and Rudd et al. [6]. Except for the low-energy yields, the observed energy spectra of the emitted electrons collected at all angles can be explained reasonably by the binary-encounter calculations outlined in Sect. 3.2. In a crude treatment, the parameter u can be chosen to be equal to the value obtained from the binding energy by use of (3.20). For target atoms other than hydrogen, however, the theoretical spectrum (3.22) based on the Fock distribution is only an approximate expression. Indeed, it has been reported that a small improvement of the calculated spectra can be obtained by using modified values of u [6].

4. Ion Channeling and High-Energy Shadowing

Ion-induced electron emission from a crystal target is influenced by ion channeling and related effects because of the restricted impact parameters in the ion–atom collisions in the crystal. In this chapter the concept of the ion-beam shadowing effect, which is familiar for keV-energy ions, is extended to MeV or higher-energy ions and is consistently linked to ion channeling. To do this, we discuss ion–atom elastic scattering over a wide energy range of the ions in a unified manner, on a classical as well as a quantum-mechanical basis. At low ion energies the ion-beam shadowing effect is directly related to the shadow of an individual atom, while at high ion energies it essentially corresponds to ion channeling near the surface, which can generally be understood by Lindhard's standard continuum model of channeling. Such a viewpoint is essential to understand electron emission under channeling incidence conditions.

4.1 Classical Coulomb Shadow

We consider the classical Rutherford scattering of a projectile by a repulsive pure Coulomb field of a target particle [13, 24, 42]. Elastic scattering of a fully stripped ion by a nucleus at rest is a typical example. Only small-angle scattering is of interest, since this is the main type of scattering responsible for the Coulomb shadow. Also, it is assumed that the projectile is in a nonrelativistic velocity range. For the relative motion of the projectile, in a coordinate system fixed to the target particle, the scattering angle ϑ of the projectile for an impact parameter b is given by

$$\vartheta = \frac{2Z_1Z_2e^2}{M_\mathrm{r}V^2b}\,, \tag{4.1}$$

where Z_1 and Z_2 are the atomic numbers of the projectile and the target, respectively, e is the electronic charge, V is the velocity of the incident ion, and $M_\mathrm{r}=M_1M_2/(M_1+M_2)$ is the reduced mass, with M_1 and M_2 being the masses of the projectile and target, respectively. The scattering angle in the laboratory frame ϑ_1 is related to ϑ by

$$\vartheta_1 = \frac{M_2\vartheta}{M_1+M_2}\,. \tag{4.2}$$

In the laboratory frame we formally use V_1 instead of V, although they are equal when the target particle is initially at rest.[1] Substituting (4.1) into (4.2), we obtain

$$\vartheta_1 = \frac{2Z_1Z_2e^2}{M_1V_1^2b} = \frac{Z_1Z_2e^2}{E_1b}, \tag{4.3}$$

where $E_1 = M_1V_1^2/2$ is the kinetic energy of the projectile. Equation (4.3) is independent of M_2 and corresponds to the limiting case of $M_2 \to \infty$, i.e. $M_r = M_1$ in (4.1), indicating the recoilless character of the small-angle Coulomb scattering in the laboratory frame.

When a parallel beam of the projectiles is incident on a repulsive Coulomb field, the deflected trajectories of the projectiles give rise to a shadow cone that the projectiles do not enter, as shown in Fig. 4.1. Using the parameters shown in Fig. 4.1, the position of the scattered projectile y far from the center of the Coulomb field ($y = z = 0$) is given by

$$y = b + z\vartheta_1 = b + \frac{Z_1Z_2e^2z}{E_1b}. \tag{4.4}$$

The radius R of the shadow cone at a distance z is defined by the minimum value of y as b is varied, i.e.

$$R(z) = 2\sqrt{\frac{Z_1Z_2e^2z}{E_1}}. \tag{4.5}$$

If the two values of $b = b_1, b_2$ for any given pair of values of z and y are taken into account in (4.4), as shown for the position P(z, y) in Fig. 4.1, the scattering intensity of the projectiles outside the shadow cone is given by the sum of the values of $|2\pi b\, db/2\pi y\, dy|$ for $b = b_1, b_2$. It follows that the scattering intensity of the projectiles I_+, which is normalized to the incident projectile flux, is given by [13, 43]

$$I_+(s) = \begin{cases} 0 & (s < 1) \\ \frac{1}{2}\left[(1-s^{-2})^{1/2} + (1-s^{-2})^{-1/2}\right] & (s > 1) \end{cases}, \tag{4.6}$$

where the normalized distance s is defined by

$$s = y/R(z). \tag{4.7}$$

While $I_+(s)$ approaches unity, i.e. the unshadowed value, with increasing s, it diverges to $+\infty$ at $s = 1$, i.e. at the edge of the shadow (see also Fig. 4.3). Note that the missing flux which would otherwise come within the shadow is πR^2, while the integral of the excess intensity $I_+ - 1$ from $y = R$ to ∞ in the two-dimensional y space amounts to $\pi R^2/2$. This implies a 50% compensation

[1] The relative motion is commonly introduced in the center-of-mass frame, but of course in this case it is identical to that in the laboratory frame.

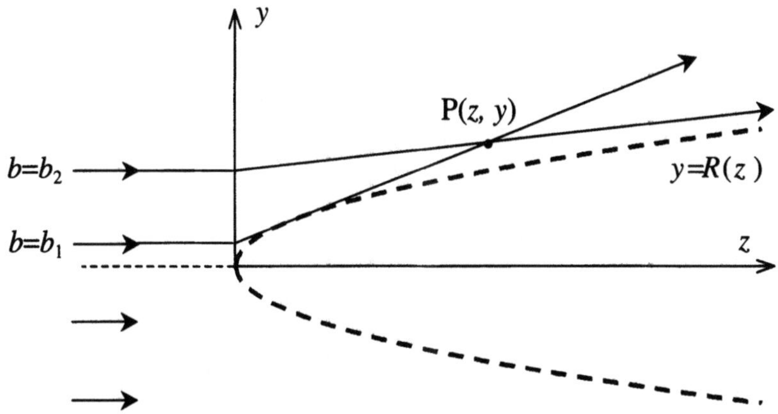

Fig. 4.1. Schematic illustration of a classical Coulomb shadow around a repulsive Coulomb field centered at $z = y = 0$. There are two trajectories ($b = b_1$ and b_2) that pass through the position P(z, y) outside the shadow cone

of the missing flux by the small-angle deflection. The compensation becomes higher in the case of a screened Coulomb potential [13].

For later discussion, the scattering by an attractive Coulomb field is briefly noted. The classical calculation of the scattering intensity I_- is similar to the case of I_+ shown above except that ϑ_1 in (4.4) is replaced by $-\vartheta_1$. The result is

$$I_-(s) = \frac{\left(s + \sqrt{s^2+1}\right)^2}{4s\sqrt{s^2+1}} \,. \tag{4.8}$$

$I_-(s)$ decreases with increasing s and approaches unity. It diverges to $+\infty$ at $s = 0$ (see also Fig. 4.5).

4.2 Quantum-Mechanical Coulomb Shadow

In the quantum-mechanical treatment a projectile of momentum $M_r V = \hbar K$, where $2\pi\hbar$ is the Planck constant, is expressed as a plane wave $\exp(\mathrm{i}Kz)$, as shown in Fig. 4.2. The wave function Φ after scattering by the Coulomb field can be written in the form [36, 37]

$$\Phi = \mathrm{e}^{\mathrm{i}Kz}\chi \,, \tag{4.9}$$

where χ is a function of the space coordinates. Φ satisfies the Schrödinger equation

$$\frac{\hbar^2}{2M_r}\nabla^2\Phi + \frac{Z_1 Z_2 e^2}{r}\Phi = \mathcal{E}\Phi \,, \tag{4.10}$$

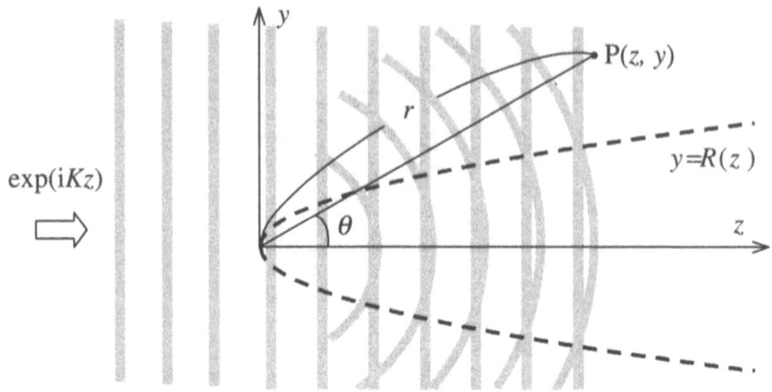

Fig. 4.2. Schematic illustration of the scattering of plane waves in a Coulomb field centered at $z = y = 0$

where $\mathcal{E} = \hbar^2 K^2/2M_\mathrm{r}$ is the kinetic energy of the projectile associated with the relative motion in the center-of-mass frame. In the polar coordinates shown in Fig. 4.2, Φ is a function of r and ϑ. In parabolic coordinates, Φ is written in terms of the parameters $\xi = r - z = r(1 - \cos\vartheta)$ and $\eta = r + z = r(1 + \cos\vartheta)$. We may anticipate that at a distance far from the center of the Coulomb field, the wave will become an outgoing spherical wave $\Phi \sim \exp(\mathrm{i}Kr)/r$, i.e. $\chi \sim \exp[\mathrm{i}K(r - z)]$, indicating that χ is inherently a function of ξ only. It follows from (4.9) and (4.10) that χ satisfies

$$\xi \frac{\mathrm{d}^2\chi}{\mathrm{d}\xi^2} + (1 - \mathrm{i}K\xi)\frac{\mathrm{d}\chi}{\mathrm{d}\xi} - \frac{\kappa K}{2}\chi = 0\,, \tag{4.11}$$

where

$$\kappa = 2Z_1 Z_2 e^2/\hbar V \tag{4.12}$$

is Bohr's Coulomb scattering parameter. The solution of (4.11) is given by the confluent hypergeometric function $F(A, B, X)$, defined by

$$X\frac{\mathrm{d}^2F}{\mathrm{d}X^2} + (B - X)\frac{\mathrm{d}F}{\mathrm{d}X} - AF = 0\,, \tag{4.13}$$

where A and B are constants. Comparison between (4.11) and (4.13) leads to

$$\chi = F\left(-\frac{\mathrm{i}\kappa}{2}, 1, \mathrm{i}K\xi\right). \tag{4.14}$$

Finally, Φ is written

$$\Phi = \exp\left(-\frac{\pi\kappa}{4} + \mathrm{i}Kz\right)\Gamma\left(1 + \frac{\mathrm{i}\kappa}{2}\right)F\left(-\frac{\mathrm{i}\kappa}{2}, 1, \mathrm{i}K\xi\right), \tag{4.15}$$

where Γ is the gamma function, defined for $\mathrm{Re}\,(z) > 0$ by

$$\Gamma(z) = \int_0^\infty \mathrm{e}^{-t} t^{z-1}\,\mathrm{d}t\,. \tag{4.16}$$

The scattering intensity I normalized to the incident projectile flux, which corresponds to (4.6) in the classical case, is given by

$$I = |\Phi|^2 = I_0 \left| F\left(-\frac{\mathrm{i}\kappa}{2}, 1, \mathrm{i}K\xi\right)\right|^2, \tag{4.17}$$

where

$$\begin{aligned} I_0 &= \exp\left(-\frac{\pi\kappa}{2}\right)\left|\Gamma\left(1+\frac{\mathrm{i}\kappa}{2}\right)\right|^2 \\ &= \pi\kappa/\left[\exp(\pi\kappa) - 1\right]\,. \end{aligned} \tag{4.18}$$

I_0 is the scattering intensity at $\xi = 0$, i.e. on the collision axis ($\vartheta = 0$). Clearly, $I_0 < 1$ for a repulsive field ($\kappa > 0$).

We are primarily interested in the forward scattering intensity in the laboratory frame, which is responsible for the Coulomb shadow. The scattering intensity is obtained simply by replacing M_r by M_1, similarly to the classical case discussed in Sect. 4.1. Correspondingly, V is formally replaced by V_1, which has been defined in the laboratory frame in Sect. 4.1. The projectile energy is therefore written

$$E_1 = M_1 V_1^2/2 = \hbar K V_1/2\,. \tag{4.19}$$

It follows that the classical shadow cone radius (4.5) can be written

$$R(z) = 2\sqrt{\kappa z/K}\,. \tag{4.20}$$

Furthermore, for forward scattering ($y \ll z$), we may write

$$\xi = \sqrt{z^2 + y^2} - z \simeq y^2/2z\,. \tag{4.21}$$

From (4.20) and (4.21), we obtain the following expression:

$$K\xi = 2\kappa\,[y/R(z)]^2 = 2\kappa s^2\,. \tag{4.22}$$

According to (4.17) and (4.22), I depends not only on s but also on κ. The dependence on the scaling parameter s is of course expected from the correspondence to the classical scattering intensity (4.6). In addition, the parameter κ is a measure of the quantum-mechanical behavior of the scattering intensity. Figure 4.3 shows $I(s, \kappa)$ as a function of s; this has been obtained numerically from (4.17) and is shown together with the classical shadow cone given by (4.6). As κ decreases, the edge of the Coulomb shadow becomes smeared as a result of the diffraction effect, so that the shadow ($s < 1$) becomes incomplete for $\kappa \lesssim 1$. The scattering intensity outside the shadow ($s > 1$) oscillates and converges to unity with increasing s. Also, the

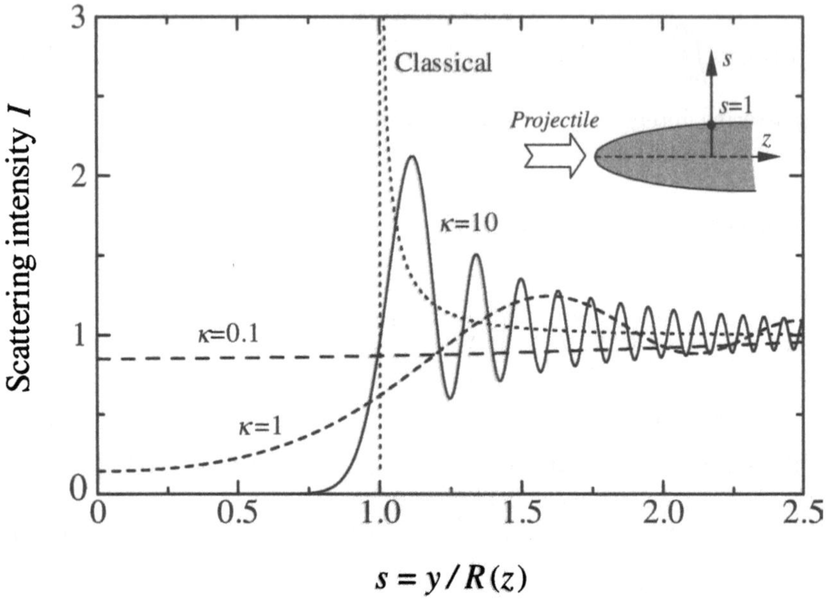

Fig. 4.3. Calculated quantum-mechanical shadow cone [44]

relative period of the oscillation (relative to the length R) becomes shorter with increasing κ, indicating that an effective intensity for $\kappa \gg 1$ can be obtained by smoothing the rapid oscillations. It is clear that $I(s,\kappa)$ becomes classical for $\kappa \gg 1$, which is the condition required for a classical treatment of the shadow cone.

The deviation from the classical shadow cone can be investigated further with the integrated intensity $\mathcal{D}(\kappa)$ given by [44],

$$\mathcal{D}(\kappa) = \frac{1}{\pi}\int_0^1 I(s,\kappa)\, 2\pi s\, \mathrm{d}s \, . \tag{4.23}$$

The calculated curve of $\mathcal{D}(\kappa)$ is shown in Fig. 4.4. $\mathcal{D}(\kappa)$ represents the averaged ion flux for $s \leq 1$ and, accordingly, is a measure of the diffraction effect, i.e. $\mathcal{D} = 1$ in the quantum-mechanical limit ($\kappa = 0$) and $\mathcal{D} = 0$ in the classical limit ($\kappa = \infty$). It is noteworthy that $\mathcal{D} = 0.86$, 0.35, and 0.11 at the representative values of $\kappa = 0.1$, 1, and 10, respectively.

The present discussion also holds for an attractive Coulomb field, for which $\kappa < 0$. We see from (4.18) that $I_0 > 1$ for the attractive Coulomb case, which is consistent with a classical picture, i.e. focusing of the incident flux by the attractive Coulomb field. It is instructive to investigate the scattering intensity for negative values of κ, which corresponds to the case of an electron or antiproton beam. Figure 4.5 shows $I(s,\kappa)$ for $\kappa < 0$ as a function of s; this has also been obtained from (4.17) and is shown together with the

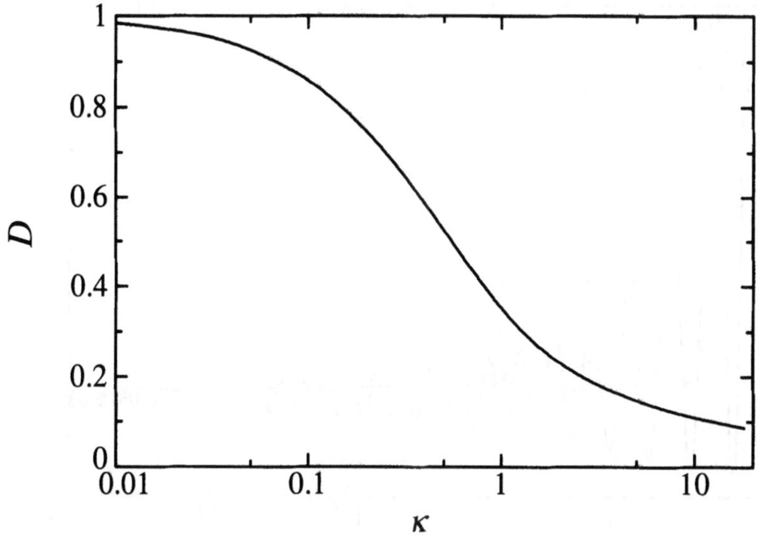

Fig. 4.4. Integrated intensity $\mathcal{D}(\kappa)$. Adapted from [44]

classical *reversed shadow cone* given by (4.8). Again, we can see the quantum-mechanical behavior of the scattered intensity for small values of $|\kappa|$, and the correspondence of the classical and the quantum-mechanical cases at large $|\kappa|$ values as well.

It is worth noting that the classical limit $|\kappa| \gg 1$ for both the repulsive and the attractive Coulomb field is approached with decreasing V. The parameter κ defined by (4.12) can also be expressed as

$$\kappa = 2\pi\rho_c/\lambda_B \, , \tag{4.24}$$

where $\rho_c = 2Z_1Z_2e^2/M_1V_1^2$ is the classical collision diameter in the recoilless limit ($M_r \simeq M_1$), and $\lambda_B = h/M_1V_1$ is the de Broglie wavelength of the ion. Briefly, (4.24) represents the ratio of the small scattering angle to the corresponding angular spread due to the diffraction of the de Broglie waves [13, 15, 42]. The former is proportional to ρ_c, according to (4.3), while the latter is proportional to λ_B. The classical picture is valid when the scattering angle is much greater than the angular spread. As V_1 decreases, ρ_c increases more rapidly than λ_B, and the scattering becomes classical.

This is contrast to the case of a potential that has a finite range L_0, such as a perfectly rigid sphere of radius L_0. In this case, the classical limit is aproached when the wavelength is much shorter than the size L_0 of the potential, i.e. for large V_1 [37]. Roughly, a screened Coulomb potential can be regarded as an intermediate case, so that the question may arise as to whether small or large V_1 corresponds to the classical limit in this case. Lindhard pointed out that the condition for the classical limit for a screened

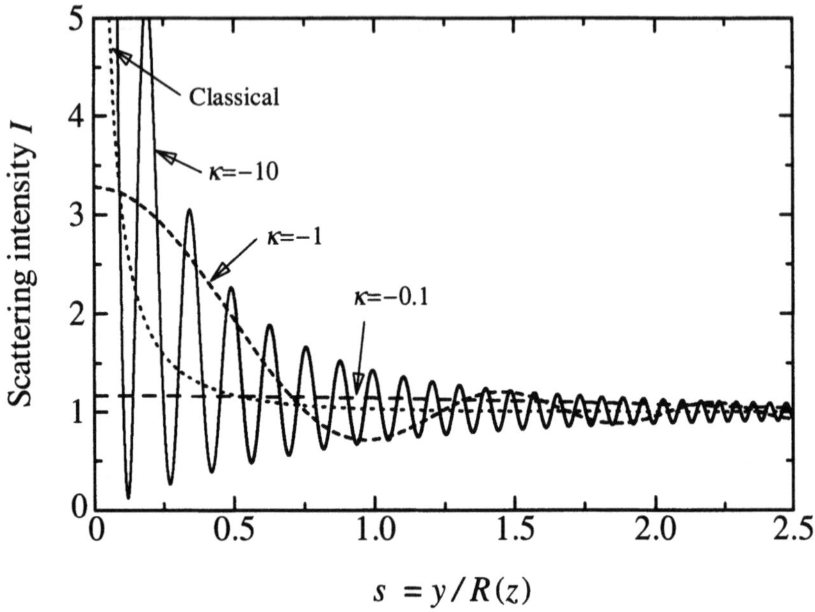

Fig. 4.5. Calculated *reverse* Coulomb shadow for negative κ

Coulomb potential of an atom is essentially similar to the unscreened case, i.e. the condition requires small V_1 [13].

4.3 Shadowing Effect: Interference of Shadow Cones

We first consider an isolated row of equally spaced atoms with a spacing d, and discuss the condition for interference of the shadow cones when the projectiles are incident obliquely on the row of atoms, as shown in Fig. 4.6. In low-energy ion-scattering spectroscopy the aligned atoms on the surface are of primary interest [45, 46], while for high-energy ions the atoms in the bulk also play significant roles. We are mainly concerned with the classical case, e.g. $\kappa \gtrsim 10$, which applies in most of the shadowing experiments. We also assume that the ion energy is not so high that the shadow cone radius at a downstream distance $z = d$ is greater than the possible thermal displacements of the atoms (~0.1 Å).

As the incidence angle ψ with respect to the row of atoms decreases, there exists a critical value of ψ at which the edge of the shadow of an atom passes through the adjacent atom. In this case, the electrostatic field of the adjacent atom should certainly perturb the trajectories of the approaching ions. Nevertheless, we may obtain roughly the critical condition for the interference of shadow cones by using the unperturbed shadow cone. Under the assumption

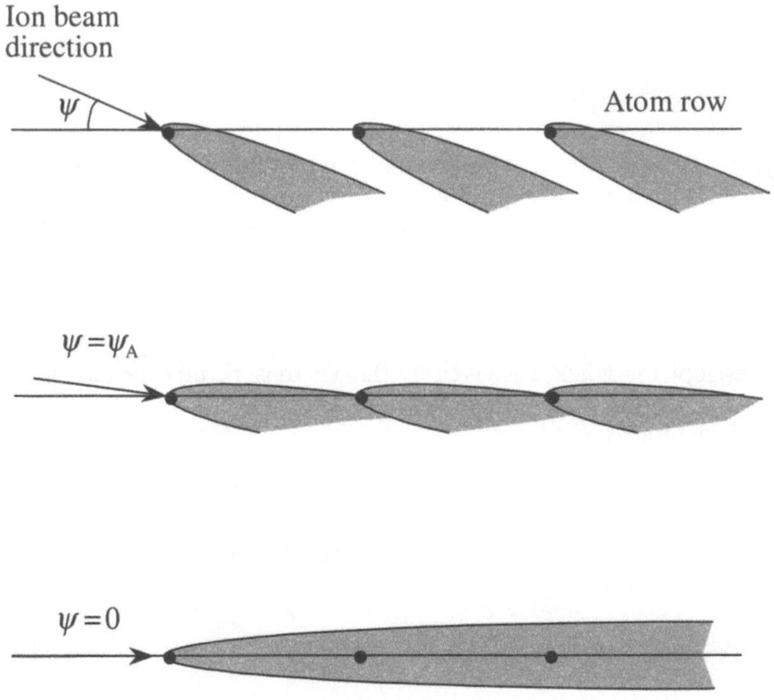

Fig. 4.6. Interference of shadow cones (schematic)

that the shadow cone results from a pure Coulomb field, a critical angle ψ_A can be defined as[2]

$$\psi_A = \frac{R_c}{d} = 2\sqrt{\frac{Z_1 Z_2 e^2}{E_1 d}}, \tag{4.25}$$

where R_c is the shadow cone radius at $z = d$ in (4.5), i.e.

$$R_c = 2\sqrt{\frac{Z_1 Z_2 e^2 d}{E_1}} = \sqrt{\frac{8 Z_1 Z_2 e^2 d}{M_1 V_1^2}}. \tag{4.26}$$

For $\psi > \psi_A$, the shadow cones of the atoms are independent each other, i.e. they are unaffected by the other shadow cones. For $\psi \leq \psi_A$, interference of the shadow cones should occur so that the atoms are effectively shadowed and are not exposed to the projectiles. The interference of shadow cones is a key concept in surface structure analysis by low-energy ion-scattering spectroscopy [45, 47].

[2] The criterion (4.25) is essentially the same as that given in Lindhard's paper ([13], p. 16).

In a realistic treatment of the ion–atom interaction, it is better to use an atomic potential of the screened Coulomb type,

$$\mathcal{V}(r) = \frac{Z_1 Z_2 e^2}{r} f_s(r/a) , \tag{4.27}$$

where f_s (≤ 1) is a screening function, and a represents the screening length [48]. One of the widely used expressions for a screened potential is the Molière potential, an approximation to the Thomas–Fermi potential, in which f_s is represented as

$$f_s(r/a) = 0.35\,\mathrm{e}^{-0.3r/a} + 0.55\,\mathrm{e}^{-1.2r/a} + 0.1\,\mathrm{e}^{-6r/a} . \tag{4.28}$$

For fully stripped fast ions a is given by the Thomas–Fermi screening length

$$a = 0.8853\, a_0 Z_2^{-1/3} , \tag{4.29}$$

where $a_0 = 0.529$ Å is the Bohr radius. For partially stripped ions an approximate expression that can be used in place of (4.29) is the Firsov screening length, which was proposed for an interatomic potential (between two neutral atoms) and is given by [49]

$$a = 0.8853\, a_0 \left(Z_1^{1/2} + Z_2^{1/2}\right)^{-2/3} . \tag{4.30}$$

Another expression used for the partially stripped case is given by [42]

$$a = 0.8853\, a_0 \left(Z_1^{2/3} + Z_2^{2/3}\right)^{-1/2} . \tag{4.31}$$

The difference between the two expressions (4.30) and (4.31) causes only minor difference in practical estimates in the ion–atom interaction.

The shadow cone radius for a screened potential is generally smaller than the value calculated from (4.5), and the difference between the two becomes smaller for higher ion energies [43,50]. The shadow cone for a screened Coulomb potential is less classical than in the unscreened case [13], as can be anticipated from the effective decrease in the value of Z_2 in the Bohr parameter (4.12).

For an incidence angle smaller than the critical angle discussed above, most of the crystal atoms near the surface are effectively shadowed. In low-energy ion-scattering spectroscopy using ions in the keV energy range, a typical critical angle is of the order of 10°. In this case, the shadowing effect is sensitive to the atomic structure of the surface since the shadow of an individual atom is responsible for the shadowing effect. The low-energy shadowing effect can therefore be applied to the structural analysis of surfaces [45,47,51].

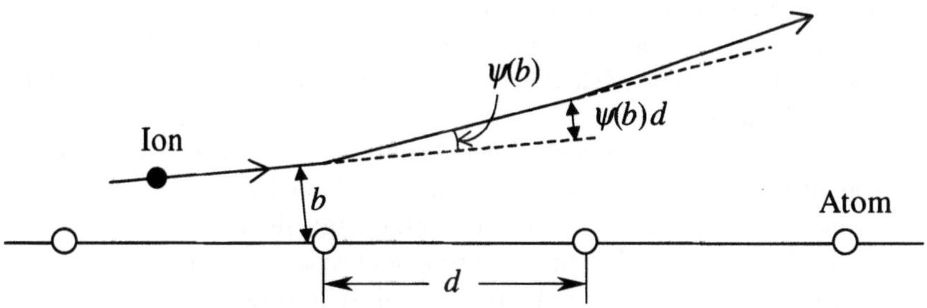

Fig. 4.7. Ion scattering by aligned atoms [44]

4.4 Shadowing Effect of High-Energy Ions

The concept of interference of shadow cones must be modified as the ion energy is increased to the MeV/u or higher range. The corresponding phenomenon at high energies is ion channeling. In fact, ion channeling can also be characterized by R_c, given by (4.26). In collisions of the ion with aligned atoms, as shown in Fig. 4.7, each soft collision, of impact parameter b, causes an increase in the impact parameter for the next soft collision by $\psi(b)d$, where $\psi(b)$ is the scattering angle in the laboratory frame. For the screened Coulomb potential described in Sect. 4.3, $\psi(b)d$ can be expressed as

$$\psi(b)d = \frac{Z_1 Z_2 e^2 g_s(b/a)\, d}{E_1 b} = \frac{R_c^2\, g_s(b/a)}{4b}, \tag{4.32}$$

where $g_s(b/a)$, a function of b/a, represents a correction factor to the case of pure Coulomb scattering [$g_s(b/a) = 1$]. This indicates that for a given target crystal the successive impact-parameter increase, i.e. the channeling effect, is characterized by the parameter R_c.

As the ion energy increases, the shadow of an atom becomes so narrow that the thermally displaced crystal atoms hardly ever remain inside the shadow cones of the adjacent atoms in the row. In this case, the shadow cone loses its reality, in contrast to the case of low-energy ions, although R_c still remains a key parameter, as noted above. Instead, for high-energy ions, an effective shadow develops along a low-index axial direction of the crystal. The effective shadow resulting from the correlated small-angle scatterings can be interpreted as an incomplete interference of the shadow cones, which is disturbed more or less by the thermal displacements of the atoms.

Similarly, at high energies a planar shadow is developed along a low-index planar direction. However, this type of shadowing may be regarded as a second-order effect which eventually disappears as the ion energy decreases to the keV energy range, unlike in the axial case. It is therefore less meaningful to describe planar shadowing in terms of the interference of shadow cones.

Both for axial and for planar shadowing at high energies, most of the crystal atoms in the surface layer are effectively shadowed against the incident ions. To specify this type of shadowing, the term *high-energy shadowing effect* is used in this book.

Obviously, the high-energy shadowing effect corresponds to ion channeling near the surface, which can be successfully explained by the standard continuum model of Lindhard. The term "high-energy shadowing", rather than "channeling", is used preferentially throughout this book since in the study of electron emission from crystals we are mainly interested in the shadowed or unshadowed target atoms, rather than in the behavior of the channeled ions.

4.5 Continuum Model of Channeling

Under high-energy shadowing conditions, most of the incident ions are channeled without suffering large-angle scattering. The motion of a channeled ion can be described well by the standard continuum model of Lindhard [13]. Since this model has been reviewed in many articles [14,15,24,43,52,53], only a few selected subjects related to this model that are of fundamental importance in studies of the high-energy shadowing effect are briefly summarized here.

4.5.1 Scaling of the Ion Trajectory

The continuum potentials which govern the lateral motion of channeled ions are obtained by averaging the periodic atomic potential along the crystal axis or over the crystal plane. Thus the continuum potential for axial channeling $U_{\mathrm{a}}(r)$ is a function of the distance r from the row of atoms, and that for planar channeling $U_{\mathrm{p}}(y)$ is a function of the distance y from the sheet of atoms. The approximate expressions given by Lindhard, which are based on a screened Coulomb potential, are

$$U_{\mathrm{a}}(r) = \frac{Z_1 Z_2 e^2}{d} \ln\left(\frac{3a^2}{r^2} + 1\right) \tag{4.33}$$

and

$$U_{\mathrm{p}}(y) = 2\pi Z_1 Z_2 e^2 a N_{\mathrm{a}} d_{\mathrm{p}} \left[\sqrt{\left(\frac{y}{a}\right)^2 + 3} - \frac{y}{a}\right], \tag{4.34}$$

where N_{a} is the atom density in the crystal, and d_{p} is the interplanar spacing. Essentially, $U_{\mathrm{a}}(r)$ increases more rapidly than $U_{\mathrm{p}}(y)$ as the distance r or y approaches 0, indicating a stronger steering force in the axial than in the planar case.

For any reasonable type of screened Coulomb potential used to describe ion–atom interactions, e.g. the Molière potential, $U_a(r)$ can be expressed in the form

$$U_{\mathrm{a}}(r) = (Z_1/d)\,G\,, \tag{4.35}$$

where G is a function of r/a. The equation of motion is given by

$$M_1 \frac{\mathrm{d}^2 \boldsymbol{r}}{\mathrm{d}t_1^2} = -\frac{Z_1}{d} \operatorname{grad} G\,, \tag{4.36}$$

where the time t_1 is related to the distance $z = V_1 t_1$ from the surface ($z = 0$) along the axial direction. We rewrite (4.36) as

$$\frac{\mathrm{d}^2 \boldsymbol{r}}{\mathrm{d}\zeta^2} = -\operatorname{grad} G\,, \tag{4.37}$$

where ζ is given by

$$\zeta = \left(Z_1/M_1 V_1^2 d\right)^{1/2} z\,. \tag{4.38}$$

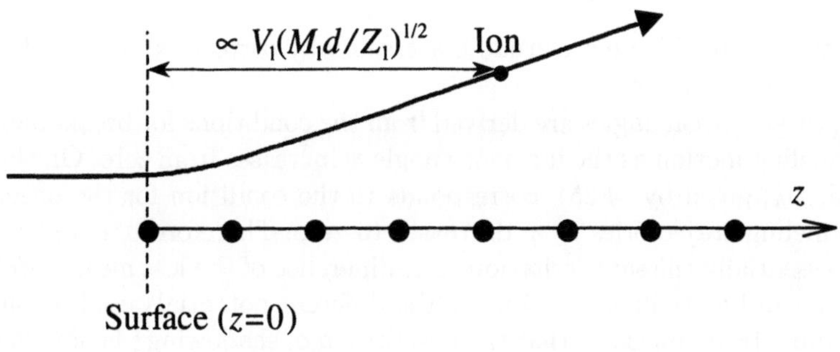

Fig. 4.8. Ion trajectory in a continuum potential field (schematic)

For fully stripped ions G does not depend on Z_1, since a is given by (4.29). It is evident from (4.37) and (4.38) that the ion trajectory $\boldsymbol{r}(z)$ in this case can be scaled with $V_1\sqrt{M_1 d/Z_1}$ with respect to the z direction, as illustrated in Fig. 4.8. Similarly, the planar trajectories can be scaled with $V_1\sqrt{M_1/Z_1 d_{\mathrm{p}}}$. For partially stripped ions, a given by (4.30) or (4.31) and, accordingly, G should depend on Z_1. The ion trajectory in this case is therefore scaled with $V_1\sqrt{M_1 d}$ and $V_1\sqrt{M_1/d_{\mathrm{p}}}$ in the axial and planar cases, respectively.

In experimental studies, it is of primary importance that, for a given axial or planar direction, the trajectories of fully stripped ions can be scaled with the projectile parameter $V_1\sqrt{M_1/Z_1}$. For example, increasing the ion velocity by a factor of 2 expands the ion trajectory along the channel by a factor of

2. Also, the trajectory of a deuteron is given by enlarging the trajectory of an equal-velocity proton along the channel by a factor of $\sqrt{2}$.

It is worth noting that the inverse of R_c as given by (4.26) is proportional to the trajectory scaling parameter $V_1\sqrt{M_1/Z_1}$, indicating that R_c^{-1} can also be used for trajectory scaling with the projectile parameters. The scaling character of the ion trajectory is well confirmed by numerical simulations, to be discussed in Sect. 4.6, and is of practical importance in the application of the high-energy shadowing effect.

4.5.2 Critical Angles

The continuum description is valid for an incidence angle less than a characteristic angle given by

$$\psi_1 = \sqrt{\frac{2Z_1Z_2e^2}{E_1 d}} \tag{4.39}$$

for axial channeling and

$$\psi_p = \sqrt{\frac{2\pi Z_1 Z_2 e^2 a N_a d_p}{E_1}}\,, \tag{4.40}$$

for planar channeling. These are crude but essentially correct expressions for the channeling critical angles.

Briefly, these critical angles are derived from the conditions for breakdown of the channeling motion as the incidence angle ψ increases from zero. On the other hand, ψ_A, given by (4.25), corresponds to the condition for the onset of the channeling trajectories as ψ decreases to zero. Therefore, ψ_A and ψ_1 stem from essentially the same criterion. The difference of the less meaningful factor of $\sqrt{2}$ can be attributed mainly to the different potentials used in the models. It must be emphasized that the description of shadowing phenomena in terms of the interference of shadow cones is physically equivalent to the continuum description of channeling, as demonstrated above for the axial critical angle.

For more detailed discussion of the high-energy shadowing effect, thermal displacements of the target atoms must be considered. Indeed, they considerably affect the distance of closest approach to a row or sheet of atoms, which is directly related to the critical angle for shadowing. Nevertheless, the critical angles (4.39) and (4.40) are rough but useful measures of the half-widths of channeling dips even in the presence of thermal displacements of the crystal atoms. In refined studies of critical angles, Barrett obtained semiempirical expressions for the critical angles on the basis of numerical calculations taking into account the thermal vibrations of the crystal atoms [43, 54], which reproduce well the observed angular half-widths of channeling dips for close-encounter events, for example ion-backscattering yields.

4.5.3 Classical Behavior of Channeled Particles

The continuum model predicts that the motion of channeled ions can be described by classical mechanics even if the Coulomb scattering parameter κ for the ion and the target atom is in the quantum-mechanical range ($\kappa \lesssim 1$) [13, 15]. Lindhard transferred the use of the Coulomb scattering parameter (4.24) to the transverse motion of channeled ions by replacing ρ_c by the distance of closest approach ($\sim a$), and λ by $\lambda_\perp = h/M_1 V_\perp$, where $V_\perp = V_1 \psi_1$ is the ion velocity in the lateral direction. The final expression for the Coulomb scattering parameter $\kappa_\perp$ for a gentle collision with an atom row is

$$\kappa_\perp = \left[4(M_1/m_e) Z_1 Z_2^{1/3} (a_0/d)\right]^{1/2} . \tag{4.41}$$

Mainly because of the factor M_1/m_e in (4.41), the classical condition $\kappa_\perp \gg 1$ is always satisfied by any ion species. Similar estimates apply in the planar case.

Accordingly, the wave packet for a channeled ion is not smeared after collision with a row or sheet of atoms, implying that the classical trajectory of the ion can be well defined. No quantum-mechanical tunneling through the center of the row or sheet of atoms occurs, even when $\kappa \lesssim 1$ for the individual scatterings. Evidently, the classical behavior thus anticipated for channeled ions justifies the numerical calculations of channeling and related effects based on the compilation of classical trajectories, even in the case of $\kappa \lesssim 1$.

4.5.4 Statistical-Equilibrium Flux Distribution

Under channeling incidence conditions, the flux density of the ions within the channel generally depends on the depth from the surface. For planar channeling, for example, a periodic change of the flux density as a function of depth is often reflected in ion-backscattering spectra [15]. With increasing depth, however, the phases of the oscillatory motions are expected to become uncorrelated each other, resulting in a flux distribution that is effectively independent of depth, or time [13]. Such a steady state of the channeled ion flux could be approximated by that given by the statistical-equilibrium distribution, i.e. the microcanonical distribution with respect to position $\boldsymbol{r}$ and momentum $\boldsymbol{p}$, for one- or two-dimensional motion, corresponding to planar or axial channeling, respectively.

For motion of a particle of total energy E_t in an n-dimensional ($n = 1, 2$) potential field $U_n(\boldsymbol{r})$, the statistical-equilibrium spatial distribution $\mathcal{F}_n \mathrm{d}^n r$ can be written

$$\mathcal{F}_n(\boldsymbol{r})\, \mathrm{d}^n r = \left(\int \delta\left[p^2/2M_1 + U_n(r) - E_t\right] \mathrm{d}^n p \right) \mathrm{d}^n r\, , \tag{4.42}$$

where δ represents the delta function, and the minimum value of U_n is taken to be zero (in most cases, at the center of the crystal channel). The simple integral in (4.42) leads to

$$\mathcal{F}_n(\boldsymbol{r})\,\mathrm{d}^n r = C_n\,[E_\mathrm{t} - U_n(\boldsymbol{r})]^{(n-2)/2}\,\mathrm{d}^n r\,, \tag{4.43}$$

where C_n is a normalization constant independent of $\boldsymbol{r}$. For planar channeling ($n = 1$), the ion flux $\mathcal{F}_1$ is enhanced near the boundary of the accessible width; this boundary satisfies $U_1 = E_\mathrm{t}$, which corresponds to $U_\mathrm{p}(y) = E_\perp$. For axial channeling ($n = 2$), the ion flux $\mathcal{F}_2$ is uniform over the accessible area where $U_2(\boldsymbol{r}) \le E_\mathrm{t}$, corresponding to $U_\mathrm{a}(r) \le E_\perp$. Clearly, a uniform flux distribution is characteristic of a two-dimensional motion.

The assumption of statistical equilibrium is often used to study the interaction of channeled ions with lattice defects. The use of the statistical-equilibrium flux simplifies the averaging process over the initial conditions of the transverse motion when channeled-ion interaction with extended lattice defects is discussed [43,55,56]. Obviously, the statistical-equilibrium approach must be applied carefully to the surface layer, where in-phase trajectories still remain [57,58].

4.6 Numerical Simulations

The trajectories of fast ions in a crystal under channeling incidence conditions can be calculated numerically under the assumption of successive binary collisions, as in the approach originally developed by Barrett [54]. The flux distribution of ions around an atom row calculated in this manner provides the depth dependence of the event of interest, such as nuclear encounters, K-shell ionization, and encounters with L-shell electrons. The method of calculation is essentially similar to the methods commonly used in ion beam analysis of surface layers [43]. An outline of the calculations is described below.

In the calculations, the thermal displacements of atoms are taken into account in the form of the two-dimensional root-mean-square amplitude of the thermal vibrations of atoms perpendicular to the axial direction, $\varrho_\perp$, which is related to the Debye temperature of the crystal. The values of $\varrho_\perp$ and the bulk Debye temperatures are listed in Gemmell's review of channeling [15].[3] Commonly, the temperature dependence of $\varrho_\perp$ can be estimated by using the Debye model. The distribution of the lateral displacement of an atom, $\mathcal{P}_\perp$, as a function of the lateral distance $r_\perp$ from the equilibrium site is assumed to be a Gaussian function:

$$\mathcal{P}_\perp(r_\perp)\,\mathrm{d}^2 r_\perp = (\pi\varrho_\perp^2)^{-1}\exp\left[-(r_\perp/\varrho_\perp)^2\right]\,\mathrm{d}^2 r_\perp\,. \tag{4.44}$$

Many sets of displacements of the atoms in the row are generated by a Monte Carlo method so that the histogram of the displacements is identical to (4.44).

[3] For surface-sensitive analyses, the surface Debye temperature, corresponding to enhanced thermal displacements of surface atoms, should be employed.

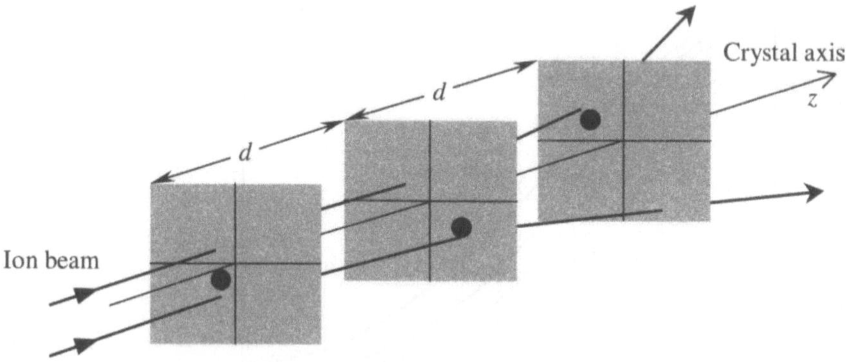

Fig. 4.9. Schematic diagram illustrating the numerical simulation of ion channeling. The atoms displaced thermally perpendicular to the crystal axis are shown by *filled circles*

For a set of displacements, shown schematically in Fig. 4.9, a number of trajectories are calculated for a given initial condition, for example parallel or grazing incidence on the row of atoms. In this case, it is usual to adopt a multistring code, in which the crystallographic symmetry of the channel is taken into account by setting suitable boundary conditions on the channel. After this process has been repeated for many sets of displacements, the flux density of the ions J can be obtained as a function of $r_\perp$ and of the depth from the surface given by $z = nd$ $(n = 0, 1, 2, ...)$. By using the depth-dependent flux J obtained in this manner, the probability $\mathcal{Q}$ for K-shell ionization, for example, of the aligned atoms as a function of z is obtained from [22]

$$\mathcal{Q}(z) = C_0 \int J(z, \boldsymbol{r}_\perp)\, \mathrm{d}^2 r_\perp \int \mathcal{K}(b) \mathcal{P}(|\boldsymbol{r}_\perp - \boldsymbol{b}|)\, \mathrm{d}^2 b\,, \tag{4.45}$$

where $\mathcal{K}(b)$ is the K-shell ionization probability as a function of the impact parameter b, and C_0 is a normalization constant, which is normally determined so that $Q = 1$ for a surface atom $(z = 0)$. Theoretical values of $\mathcal{K}(b)$ are obtained from calculations based either on a semiclassical Coulomb approximation [59] or on a binary-encounter approximation [60]. The difference between the two typically leads to only minor differences in the calculated $\mathcal{Q}(z)$. Figure 4.10 shows Si K-shell ionization probabilities for 18 and 24 MeV He^{2+} incident in a Si$\langle 110 \rangle$ direction at room temperature ($\sim$295 K), calculated using values of $\mathcal{K}(b)$ given by a semiclassical Coulomb approximation. The calculated nuclear-encounter probability for 18 MeV He^{2+} is also shown for comparison.

When the shadowing effects for other physical events are of interest, only $\mathcal{K}(b)$ has to be replaced in (4.45) by an appropriate expression for, for example, the dependence of L- or M-shell ionization on b [59]. Obviously, for

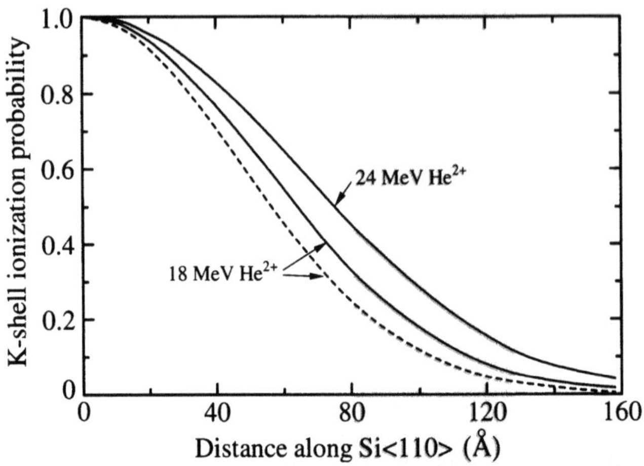

Fig. 4.10. Si K-shell ionization probabilities calculated for 18 and 24 MeV He^{2+} incident in a $\langle 110 \rangle$ direction on Si at room temperature. The *dashed curve* shows the nuclear-encounter probability for 18 MeV He^{2+}. Note that the probability values were obtained at intervals of 3.84 Å, the $\langle 110 \rangle$ interatomic spacing [66]

calculation of the nuclear-encounter probabilities, we may assume a delta function, i.e. $\mathcal{K}(b) = \delta(b)$.

Generally, an axial direction of a crystal corresponds to an intersection of several crystal planes. Therefore, calculations based on a multistring model of axial shadowing can automatically simulate planar shadowing if the direction of incidence of the ions is changed slightly from an axial to a planar direction. Furthermore, numerical calculations demonstrate well the scaling of the ion trajectories with respect to various parameters, as discussed in Sect. 4.5.1, even when thermal displacements of the atoms are taken into account.

It must be pointed out that the numerical method presented here is based on (4.44), i.e. on uncorrelated thermal displacements of crystal atoms. More realistic calculations taking correlated thermal displacements into account have been reported by Barrett and Jackson [61] and Jackson et al. [62]. In these cases, the presence of correlation typically reduces the calculated surface-peak yields of backscattered ions by a factor of 0.8–0.9 relative to the uncorrelated case.

4.7 Dechanneling

In general, the channeling trajectories of ions undergo perturbations owing to multiple scattering by electrons and thermal displacements of crystal atoms. These perturbations increase the transverse energies of the ions, which causes transitions from channeled to nonchanneled trajectories. Lattice defects and impurities in a crystal also give rise to channeling-to-nonchanneling transitions. These processes, referred to as dechanneling, are certainly expected to degrade the high-energy shadowing effect.

Under most experimental conditions, however, the intrinsic dechanneling in a defect-free crystal has a negligible effect in the observation of the high-energy shadowing effect using ion-induced electrons. This is mainly due to the finite escape length of the electrons, to be discussed in Sect. 8.4.5. In this sense, ion-induced electron emission is sensitive to the lattice structure near the crystal surface.

The influence of lattice defects on channeled trajectories has been fully studied by many workers using Rutherford backscattering spectroscopy [43, 57, 63]. In a crystal containing lattice defects, the displaced atoms enhance the effective number of unshadowed atoms. Moreover, the ions dechanneled by the lattice defects behave like random-incidence ions. The former and latter correspond to the so-called direct-scattering and dechanneling processes, respectively, in Rutherford backscattering spectroscopy. Obviously, these two factors increase the electron yield measured in the backward direction, so that both of them must be taken into account in high-energy-shadowing analysis of a crystal with defects using ion-induced electrons.

5. Experimental Methods

Experimental methods associated with the ion-induced electron spectroscopy of crystalline targets are presented in this chapter. These methods are the combination of the surface analysis techniques of electron spectroscopy and ion backscattering spectroscopy.

5.1 Electron Energy Analysis

Electrostatic energy analysis is commonly applied to electrons of kinetic energies lower than ~10 keV, with which we are mainly concerned. The principle and operation of electrostatic spectrometers for electron energy analysis are well established [64, 65]. In the early work on ion-induced-electron spectroscopy combined with ion channeling, a conventional cylindrical-mirror analyzer [20, 21] or a single-stage parallel-plate spectrometer [22] was adopted. Later, the author's group used electron spectrometers designed to satisfy particular requirements for angular and energy acceptance in their work on ion-induced-electron spectroscopy [66,67,70]. In the following, details of such spectrometers and related instruments are described.

5.1.1 Parallel-Plate Spectrometer

Figure 5.1 shows a 45° parallel-plate spectrometer of the double-deflection type designed for energy analysis of electrons at a backward angle of 180° with respect to the ion beam direction. This type of spectrometer has been used mainly for measurements of Auger electrons emitted from crystal targets. The relation between the kinetic energy of the analyzed electrons E and the plate voltage V_p is given by

$$E\,(\text{keV}) = 1.5 \times V_p\,(\text{kV})\,. \tag{5.1}$$

The angular accceptance is about 3×10^{-4} sr, which is small enough for angle-resolved electron spectroscopy, as will be discussed in Sect. 7.1.2. The electrons are energy analyzed twice by the symmetric electrostatic fields before entering the channel electron multiplier (to be described in Sect. 5.1.3). The double filtering of the measured electrons is effective in stopping stray

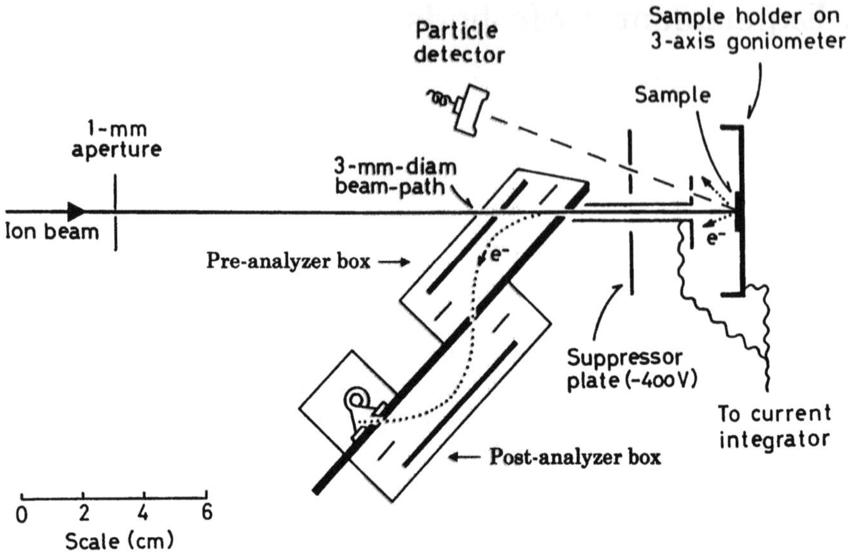

Fig. 5.1. A double-stage 45° parallel-plate spectrometer, mainly used for ion-induced Auger electron spectroscopy from crystal targets [66]

electrons from entering the electron multiplier. In many cases, a high background is observed in the electron spectrum when a single-stage parallel-plate spectrometer, which corresponds to the preanalyzer box in Fig. 5.1, is used for 180° measurements. Nevertheless, the single-stage type can be applied to electron measurements at other observation angles, under the condition that the ion beam passes outside the spectrometer and is incident on the target. For measurement of electrons of kinetic energy E, the relative energy resolution of the spectrometer $\Delta E/E$ is given by

$$\Delta E/E = \Delta S/S \, , \tag{5.2}$$

where ΔS is the width of the two identical slits placed at the entrance and exit of the postanalyzer box, which limit the electron path into the channel electron multiplier (Fig. 5.1), and S is the spacing between the two slits. For practical reasons, the relative energy resolution is adjusted to at least 1–3% to achieve a sufficient count rate of the electrons.

It is typical of deflection-type spectrometers that the energy acceptance of the spectrometer ΔE is proportional to E, as can be seen from (5.2). The electron counts in the measured spectrum are sometimes divided by E for correction of the E-dependent energy acceptance. Such a correction has to be applied in the analysis of the Auger electron yield in Sect. 7.2.

Figure 5.2 shows a 45° parallel-plate spectrometer of the mirror-symmetry type, which provides a variable energy acceptance by allowing one to adjust the slit in the spectrometer from outside the target chamber. The electrons

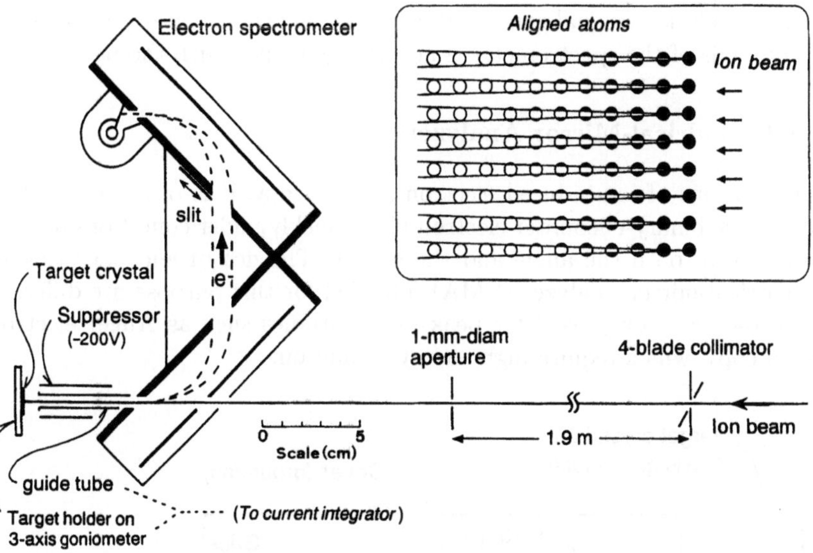

Fig. 5.2. A 45° parallel-plate spectrometer of the mirror-symmetry type, used for measurements of binary-encounter electrons emitted from crystal targets. The *inset* illustrates the high-energy shadowing effect (the *circles* schematically show the distribution of a high density of electrons) [69]

are again energy analyzed at 180° with respect to the incident beam direction over a solid angle of 2×10^{-3} sr. Equations (5.1) and (5.2) also hold for this spectrometer, except that ΔS in (5.2) is effectively given by the width of the wider slit in the postanalyzer box.

The variable energy acceptance is useful for channeling experiments since a higher count rate is preferable to a high energy resolution for quick alignment of the crystal. The variable window slit plays another important role in the experiments. When the target crystal is bombarded by fast, light ions (especially deuterons), γ rays are sometimes emitted from the target, impinge on the channel electron multiplier, and produce a uniform background in the electron energy spectrum. This background can, however, be measured on its own by closing the window slit of the spectrometer so that no electrons enter the electron multiplier, and can be subtracted from the observed spectra.

In Figs. 5.1 and 5.2, a small aluminum pipe with a disk on its end is placed between the sample and the entrance of the spectometer. This effectively shields the electron path into the spectrometer from the electric field produced by the suppressor plate. Furthermore, the beam current can be accurately measured by merging the current flowing into the pipe with that flowing into the sample.

The setup of these spectrometers needs to ensure that no stray beam strikes the interior of the spectrometer. This can be confirmed by removing the sample holder and letting the ion beam pass out of the target chamber.

Usually, no electron signal generated by any stray beam is observed when the entrance axis of the spectrometer is accurately aligned with the beam path.

5.1.2 Cylindrical-Mirror Analyzer

In applications of ion-induced electron spectroscopy, it is often required to measure the binary-encounter electron yield quickly under conditions of wide acceptance of both the angle and the energy. The design and operation of a cylindrical-mirror analyzer (CMA) intended for this purpose are different from those commonly used for peak-search studies such as Auger electron spectroscopy, which require high energy resolution.

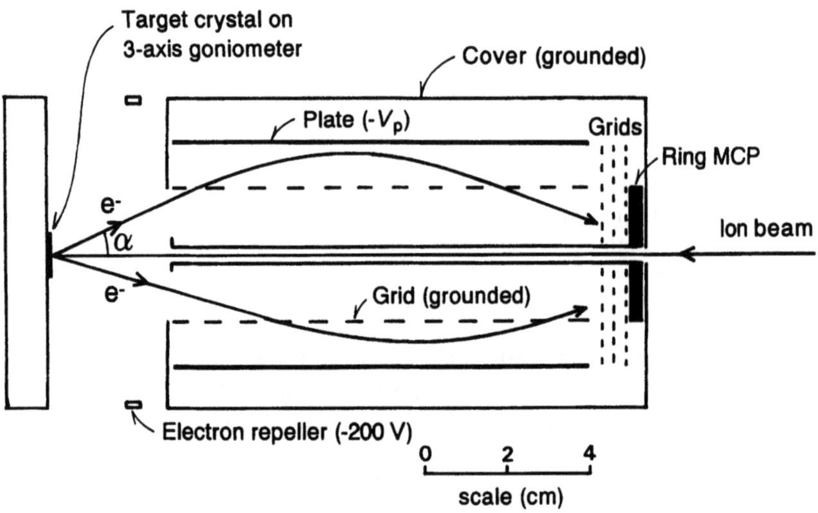

Fig. 5.3. Cross-sectional view of a cylindrical-mirror analyzer for binary-encounter electron spectroscopy [70]

Figure 5.3 is a schematic diagram illustrating a cylindrical-mirror analyzer that was developed for binary-encounter electron spectroscopy and has been used in applications of that technique. The electrically grounded grid of cylindrical shape (30 mm in diameter), fixed inside the analyzer, consists of an array of 0.2 mm diameter copper wires stretched ~4 mm apart, parallel to the cylinder axis. A microchannel plate (MCP) electron multiplier with a 4 mm diameter center hole is used for detection of the energy-analyzed electrons. The incident ion beam passes through the 3 mm diameter aluminum tube (Al thickness 80 μm) along the cylinder axis. This tube and the 7 mm diameter disk fixed at the end of the tube prevent, geometrically, the backscattered ions from entering the MCP. The three parallel grids in front of the MCP are used to set the lowest measurable electron energy. A negative voltage

V_g is applied to the center grid, while the others are electrically grounded. When the distance $\mathcal{L}$ from the entrance of the analyzer to the target crystal is 30 mm, the analyzer accepts the ion-induced electrons emitted from the crystal surface over an angular range of 7–27° with respect to the cylinder axis. The electrons are then deflected towards the cylinder axis by the electrostatic field between the cylindrical grid and the coaxial cylindrical plate, to which a negative voltage V_p is applied. The ring-shaped electron repeller (−200 V), which prevents escape of the low-energy electrons, permits accurate measurements of the ion beam current flowing into the target crystal.

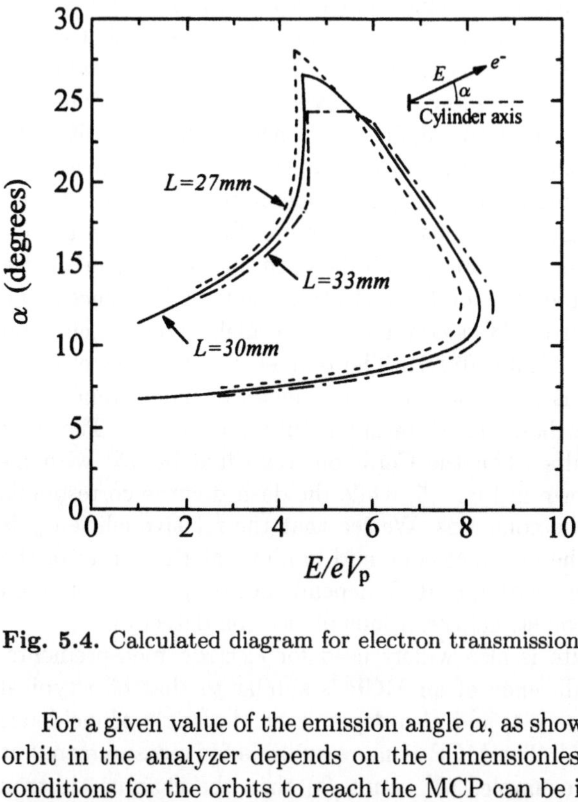

Fig. 5.4. Calculated diagram for electron transmission through a CMA [70]

For a given value of the emission angle α, as shown in Fig. 5.3, the electron orbit in the analyzer depends on the dimensionless parameter E/eV_p. The conditions for the orbits to reach the MCP can be investigated by numerical calculations assuming a logarithmic electrostatic field in the analyzer. Two calculated orbits are shown in Fig. 5.3 for the two pairs of parameters (α, E/eV_p) = (24°, 5.0) and (15°, 7.0). Figure 5.4 is a calculated diagram of the initial conditions of the electrons, shown as a function of α and E/eV_p. The results are shown for three values of $\mathcal{L}$. The electrons with initial conditions inside the area surrounded by the curve reach the MCP, while in other cases (outside this area) the electron orbit passes outside the MCP or is interrupted by the central tube. For a setup with $\mathcal{L}$ = 30 mm, a wide angular range of approximately $8° \leq \alpha \leq 25°$ with respect to the cylinder axis, which corresponds to a solid angle of 0.53 sr, is accepted for $4.5 \leq E/eV_p \leq 6.0$.

The dependence of the acceptance angle on the value of $\mathcal{L}$ can be seen from the calculated curves for $\mathcal{L} = 27$ and 33 mm, as well as for 30 mm. A slight change of $\mathcal{L}$ by 1 mm, for example, caused by a tilt of the crystal, affects the measured electron yield sensitively only at $E/eV_\mathrm{p} \simeq 8$.

5.1.3 Efficiency of Electron Multipliers

The most commonly used detector for electrons at energies lower than ~10 keV is the channel electron multiplier. Normally, the noise level can be reduced to lower than ~0.01 counts/s with careful electrical shielding of the detection system. The absolute detection efficiency of the multiplier $\mathcal{E}_\mathrm{e}$ is represented by the ratio of the number of incoming electrons to that of the true signals generated. For electrons at keV energies, the reported values of $\mathcal{E}_\mathrm{e}$ are in the range 0.3–1.0, depending on the type of channel electron multiplier and the operating conditions [6, 71, 72]. Generally, $\mathcal{E}_\mathrm{e}$ is sensitive to the surface conditions inside the electron multiplier and, for this reason, it decreases gradually with increasing integrated operation time, even during use under ultrahigh-vacuum conditions. Furthermore, $\mathcal{E}_\mathrm{e}$ depends on the incidence condition of the electrons. Figure 5.5 shows the measured dependence of the relative detection efficiency for 4 keV electrons on their incidence angle and position for the Ceratron [73], a commercially available channel electron multiplier (Murata, model EMT6081B). The diameter of the test electron beam was ~0.3 mm. In this case, the effective diameter of the entrance cone was ~8 mm and the inner diameter of the spiral tube was ~1 mm. The solid curve represents the results when the Ceratron was tilted by 45° with respect to the A–B axis shown in Fig. 5.5, while the dashed curve corresponds to incidence parallel to the cone axis. We see that the relative efficiency is maximal at the edge of the entrance cone and minimal at the center of the cone ($X_1 \simeq 0$ mm), demonstrating that $\mathcal{E}_\mathrm{e}$ depends not only on the electron energy but also on the geometrical conditions of electron detection.

The microchannel plate is also widely used for electron measurements. The electron detection efficiency of an MCP is similar to that of a typical channel electron multiplier [74] and also depends on the angle of incidence onto the MCP surface [75]. The MCP is more suitable for measurements at high counting rates (for example, 10^4 counts/s) than the channel electron multiplier, but the noise level is higher. This difference in the characteristics is one of the important factors in the choice of electron detector appropriate for any given experimental conditions. It is also to be noted that electron detectors are sensitive to γ rays emitted from the bombarded target, as mentioned in Sect. 5.1.1, which cause an apparent increase in the number of electron signals.

In ion-induced-electron spectroscopy as applied to crystal targets, there are two typical types of experiments, (i) measurements of the energy spectra under channeling and random (nonchanneling) incidence conditions, made by slightly changing the orientation of the target crystal (within ~3°), and (ii)

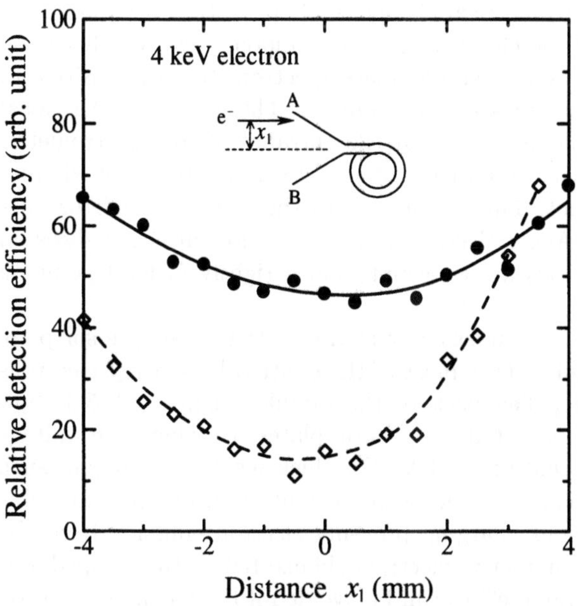

Fig. 5.5. Relative electron detection efficiency of a channel electron multiplier (Ceratron). The *solid* and *dashed curves*, drawn to guide the eye, correspond to different directions of incidence into the multiplier (see text) [73]

measurements of the electron yield at a fixed energy as a function of the angle of incidence on the target crystal. In both cases, the quantities to be discussed are the ratios of the channeling to the random yield at the same value of the electron energy. While the measured electron yield depends on the detection efficiency of the electron multiplier, the ratio of the yields is independent of the efficiency since it is canceled out. Therefore, it is unnecessary to know the absolute detection efficiency of the electron multiplier, although it is of technical importance in the experiments. This is an important feature compared with ion-induced-electron spectroscopy for the study of electron production cross sections, where the detection efficiency is of essential importance.

5.2 Experimental Arrangement

In many cases, detection of electrons at 180° with respect to the incident beam direction is advantageous for the purpose of avoiding errors associated with misalignment of the spectrometer. Once the spectrometer is accurately placed at 180°, its entrance axis always passes through the irradiated spot on the surface of the sample even when the sample is tilted. In this case, the electron count rate is unaffected by a small shift of the surface position along the beam axis, as is anticipated to occur when the sample is tilted. At other detection angles, such a geometrical change tends to disturb accurate measurements of the ratio of the channeling to the random yield. Furthermore, electron measurement by a 45° parallel-plate spectrometer is insensitive to the beam spot size on the surface of the sample. This is because the focal point of the electron paths is located at the entrance of the spectrometer, rather than on the surface, and therefore the electrons are collected from all over the irradiated spot. This is contrast to the cylindrical-mirror analyzer commonly used for Auger electron spectroscopy for surface analysis, in which the analyzed electrons are collected from a tightly focused point on the surface.

To measure the ion beam current correctly, the electron suppressor plate must be carefully designed so that most of the emitted low-energy electrons (less than ~50 eV) are repelled back to the sample. Figures 5.1–5.3 show examples of successfully operated suppressor plates, to which the voltage applied is in the range −400 to −200 V. For thick insulating samples such as alkali halide or oxide crystals, the beam current often becomes unstable because of charging. Serious charging prevents electron emission from the surface, as a result of the attractive electric field exerted by the stopped ions in the insulator. The charging effect can be avoided if the mean penetration length of the ions exceeds the thickness of the sample [44, 76]. In some cases, heating of the samples reduces the charging effect sufficiently and allows spectroscopy of ion-induced electrons [77, 78].

The ion beams used in the experiments are usually obtained from electrostatic accelerators. In most of the measurements, the beam spot size on the target is 0.1–1 mm in diameter and the beam current lies in the nanoampere range. A typical beam dose required for measurement of the high-energy shadowing effect is 10^{12}–10^{15} ions/cm^2. For most metal and semiconductor crystals, high-energy shadowing is hardly affected by beam-induced damage for such a beam dose. For some insulators, however, the beam-induced damage is serious because electronic excitations by the ions cause displacement of the crystal atoms, in contrast to the case of metals and semiconductors [43].

A knowledge of the irradiation condition of the crystal surface during the experiment must be discussed for general interpretation of the experimental data. Even for a beam dose of 10^{15} ions/cm^2 on a Si(100) surface, for example, the mean number of incident ions per Si$\langle 100 \rangle$ channel (7.4 Å^2) is only ~0.7. The irradiation conditions are similar in most of the experiments of this kind. Nevertheless, we often draw envelopes of a number of ion trajectories around an atom row to illustrate the concept of high-energy shadowing, for example, as shown in the inset in Fig. 5.2. It must therefore be recognized that the periodicity of the crystal lattice leads to the above picture, which results from the artificial superposition of the actual trajectories. These trajectories are in fact separate from each other, so that there is no correlation between them in the crystal.

5.3 Crystal Alignment

Obtaining the alignment of a given crystal axis or plane with respect to the ion beam direction is one of the fundamental techniques in channeling and related experiments [43, 57]. This operation is actually the determination of the crystallographic orientation of the solid target from the observed dependence of close-encounter phenomena, such as Rutherford backscattering or inner-shell X-ray production, on the incident direction of the ion beam. The ion-induced electron yield at a fixed energy can also be used to find the channel direction. This procedure can be carried out automatically using a computer-controlled three-axis goniometer, by recording the electron yield as a function of two tilt angles θ and ϕ of the crystal with respect to two independent rotation axes. Figure 5.6 shows an example of the data obtained during the crystal alignment of a sapphire (α-Al_2O_3) specimen using the 5–7 keV electron yield induced by 7 MeV H^+. The minimum yield at $\theta = 0.15°$, $\phi = 0.10°$ corresponds to the $\langle 0001 \rangle$ axis.

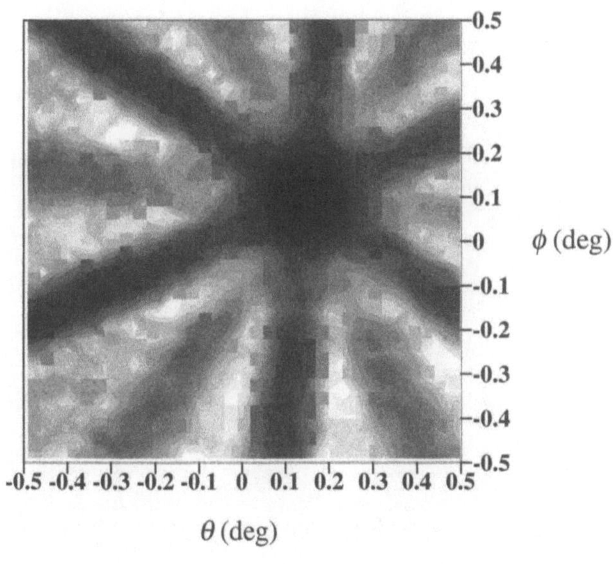

Fig. 5.6. Tilt angle dependence of the electron yield emitted from sapphire (0001) bombarded by 7 MeV H^+. Darker contrast indicates a lower electron yield [79]

To identify several sets of planar dips in the θ–ϕ mapping, the angular divergence of the beam must be much less than the critical angle given by (4.40). The beam can be tightly collimated with double apertures so that the angular divergence becomes negligible compared with the critical angle. It should be noted that the allowed angular divergence for electron measurements under channeling incidence conditions is typically less than that for ion backscattering experiments because the angular dip for the electron yield is narrower than for the backscattered ions (see the angular dips for GaP and InP shown in Sect. 10.3). In many cases, therefore, the electron yield for incidence of a zero-angular-divergence beam must be estimated by using a beam collimation technique, as will be described in Sect. 8.2.

6. Electron Escape from Solid Targets

This chapter describes experimental and numerical studies of elastic and inelastic scattering of electrons at keV energies which are of interest in ion-induced-electron spectroscopy. The discussion here is restricted to the case of noncrystalline solids since structural effects are not important for the physical processes under consideration. Electron escape from a crystal lattice is discussed in Chap. 7.

6.1 Electron Scattering in Solids

Energetic electrons produced by Auger transitions or by the binary-encounter processes discussed in Chap. 3 suffer elastic and inelastic scattering in the solid before they escape from the surface. These scattering events can also produce energetic secondary electrons in the solid. A general understanding of these processes can be obtained from electron backscattering experiments using a monoenergetic, parallel beam of electrons. Numerical simulations of electron multiple scattering in solids can also provide detailed information about the dependence of the phenomenon on the physical parameters over wide ranges. Such studies have been carried out extensively since early 1970 and have provided knowledge of the behavior of energetic electrons and positrons in solids that can be used in electron probe microanalysis of materials [80–83], for example.

The angular deflection of energetic electrons in solids results mainly from electron–atom elastic scattering. Figure 6.1 shows screened Rutherford differential cross sections of Cu for 0.1 and 1 keV electrons, calculated by Valkealahti and Nieminen [83]. The distribution of the scattering angle becomes more sharply forward-directed with increasing electron energy. It follows that Auger electrons at keV energies produced in a subsurface layer, for example, will escape from the surface without suffering large-angle deflections (see also Sect. 7.2.2). In contrast, large-angle backscattering is necessary for the escape of forward-directed keV electrons from the back surface. We recall that the binary-encounter electrons produced from target atoms are forward-directed (Chap. 3). Therefore, the most probable escape path for the electrons includes a large-angle backscattering by a crystal atom, as shown in the schematic illustration in Fig. 6.2.

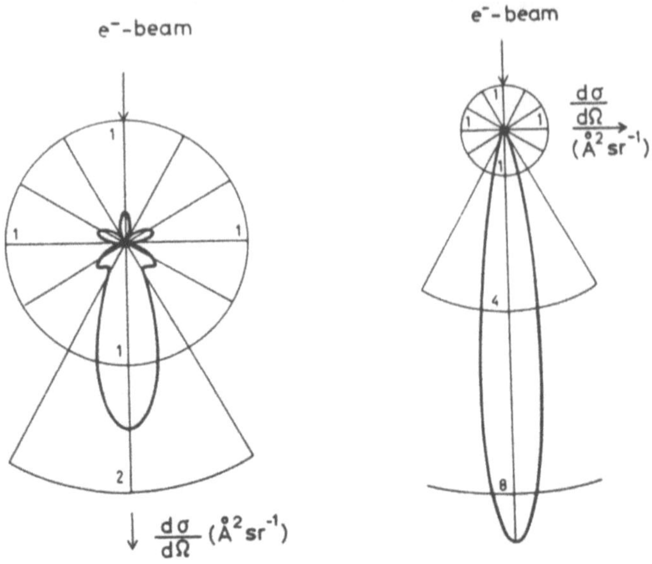

Fig. 6.1. Polar plot of the differential elastic-scattering cross sections, calculated for 0.1 keV (*left*) and 1 keV (*right*) electrons incident on a Cu atom. Adapted from [83]

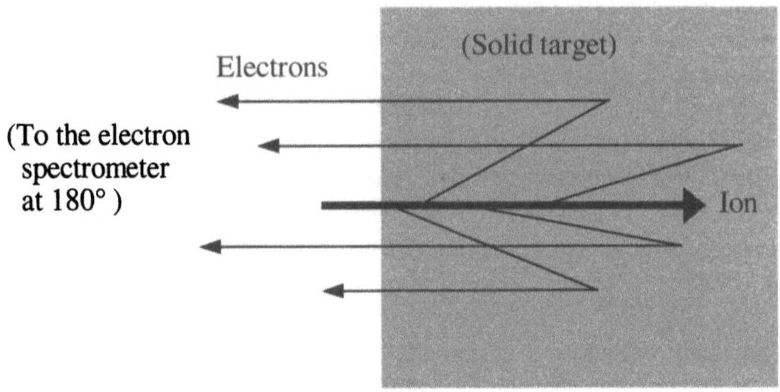

Fig. 6.2. Most probable escape paths of electrons measured at a backward angle of 180°. Adapted from [84]

In earlier work, Matsukawa et al. measured backscattering energy spectra of 10–30 keV electrons from C, Al, Au, and a Cu–Au alloy [80]. They found that the ratio w_{peak} of the most probable backscattered energy to the incident electron energy is equal to ~0.9 for Au, and w_{peak} becomes smaller for lower-atomic-number materials. These results were well reproduced by

Monte Carlo calculations of elastic and inelastic collisions of electrons in solids. Valkealahti and Nieminen calculated the slowing down of keV electrons by a Monte Carlo method based on a screened Rutherford cross section and a semi-empirical expression for electron excitations in solids. Figure 6.3 shows backscattering-energy distributions dY_B/dw as a function of the ratio of the backscattered to the incident electron energy w, calculated for 2.1 and 6.2 keV electrons incident in the normal direction on a semi-infinite plane of Cu. In this case, $w_{peak} = 0.95–1.0$ for Cu. Similarly, Valkealahti and Nieminen obtained $w_{peak} = 0.75–0.9$ for Al. These results are consistent with the experimental results of Matsukawa et al.

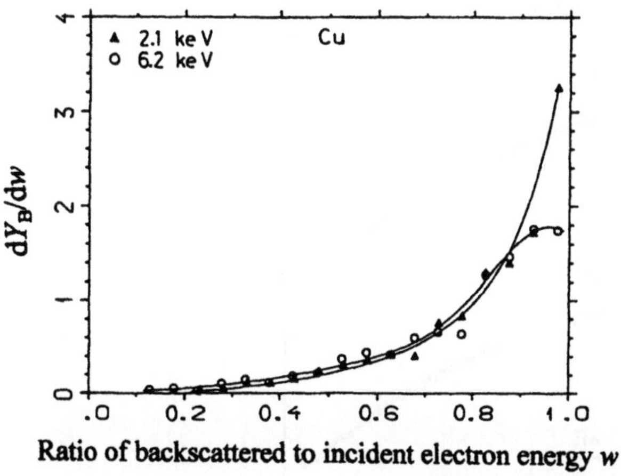

Fig. 6.3. Calculated backscattering-energy distributions of 2.1 and 6.2 keV electrons incident on a thick Cu target. Adapted from [82]

The most probable backscattered energy corresponds to the elastic scattering peak, which is generally observed in the backward direction when a solid is bombarded by monoenergetic electrons of energy lower than 1 keV [2]. Therefore, over a wide kinetic-energy range of the electrons, elastic and quasi-elastic scattering processes are predominantly responsible for the backscattered energy spectra of electrons.

6.2 180° Emission and Effective Escape Lengths

Some fundamental aspects of the backward electron emission from a solid target can be revealed by electron measurements at a backward angle of 180°. The technique of 180° electron spectroscopy provides energy spectra of unique physical character, in addition to the technical merit associated with the alignment of the spectrometer's entrance axis mentioned in Sect. 5.2. Figure

6.4 shows energy spectra of secondary electrons from GaAs under random incidence conditions, measured for oblique ($\theta_s = 45°$) and perpendicular ($\theta_s = 90°$) incidence of 3.75 MeV/u O^{8+} and S^{15+}. The electron yields are shown after they have been divided by Z_1^2 to scale of the binary-encounter yield, as discussed later (Sect. 8.1). Clearly, there is no discernible difference between the two pairs of the spectra. Similar results are shown in Fig. 6.5 for 95 keV H^+ incident on noncrystalline Ni and Au [85]. Note that for Ni the LMM Auger yield at 0.6–0.8 keV is superposed on the continuum yield.

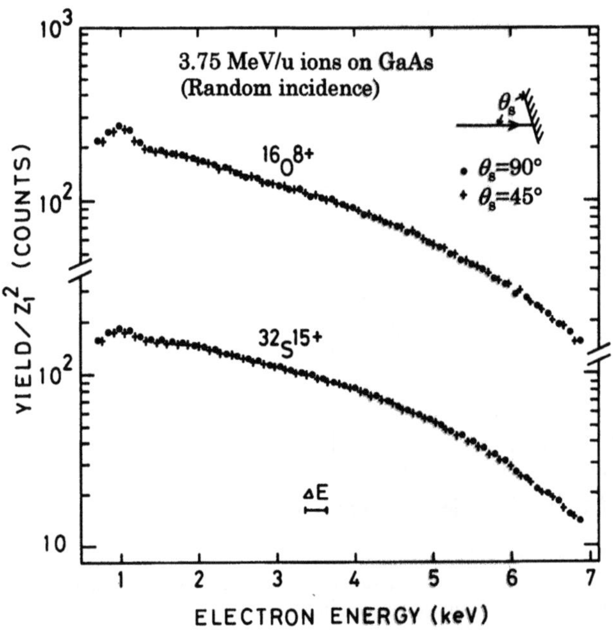

Fig. 6.4. Energy spectra of secondary electrons induced by 3.75 MeV/u O and S ions, measured under oblique incidence conditions [86]

An energy spectrum that is independent of the surface normal direction, i.e. of the sample orientation with respect to the beam direction, is characteristic of the detection at 180° of electrons that exit in a straight line. Suppose that the electrons are measured at a backward angle Υ with respect to the beam direction, as shown in Fig. 6.6. In most cases, refraction of the electrons due to the work function (at most ~5 eV) when they traverse the surface is negligible for the keV electrons under consideration. In the bombarded surface layer, we may assume a steady-state (time-independent) spatial and momentum distribution for all the ionized electrons. In this case, the momentum distribution of the electrons within a small volume element in the surface layer is a function of the distance z along the beam direction. Let ΔY be the component of the measured spectrum that is contributed by

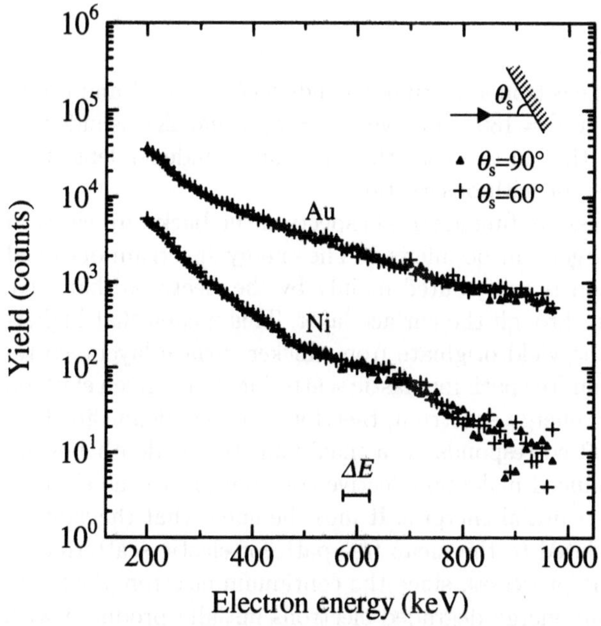

Fig. 6.5. Energy spectra of secondary electrons induced by 95 keV H^+, measured under oblique incidence conditions [85]

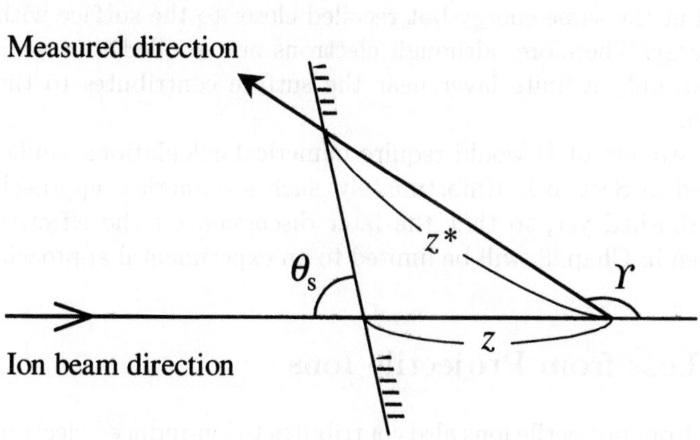

Fig. 6.6. Electron escape from a solid target

this volume element. Clearly, ΔY depends not only on z but also on z^* since the outgoing electrons from this volume element suffer energy loss along the straight escape path towards the surface. For the geometry shown in Fig. 6.6, z^* can be written

$$z^* = \frac{z \sin \theta_s}{-\sin(\varUpsilon + \theta_s)} . \tag{6.1}$$

Generally, the measured electron spectrum depends on θ_s since z^* and therefore ΔY depend on θ_s. At $\varUpsilon = 180°$, however, $z^* = z$ so that ΔY is independent of θ_s. This ensures that in this case the spectrum is independent of θ_s, which is certainly consistent with observation.

From the above discussion, fundamental aspects of the backward electron emission from a solid target can be inferred. The energy spectrum observed in the backward direction is contributed mainly by the electrons that have passed in a straight line through the surface layer. This means that higher-energy components of the yield originate from thicker surface layers, corresponding to a longer mean free path for elastic scattering. For a given electron energy in the continuum energy spectrum, therefore, we may define an effective escape length D. D corresponds to a maximum target depth beyond which the electrons produced make no effective contribution to the electron yield, irrespective of their initial energies. It must be noted that the value of D cannot be directly related to the mean free path for elastic scattering. It depends also on inelastic processes, since the continuum electron yield contains a contribution from energy-degraded electrons initially produced with higher energies than the observed energy. The deeper the electrons are recoiled in the solid, the larger is the energy loss suffered before escape. If the depth of recoil is too large, the electrons will lose so much energy that their number becomes negligible compared with the much larger number of electrons detected at the same energy but recoiled closer to the surface with a lower initial energy. Therefore, although electrons are recoiled in a wide region of the solid, only a finite layer near the surface contributes to the observed spectrum.

A theoretical estimate of D would require numerical calculations similar to those mentioned in Sect. 6.1. Unfortunately, such a numerical approach has not been established yet, so that the later discussion on the effective escape length, given in Chap. 8, will be limited to an experimental approach.

6.3 Electron Loss from Projectile Ions

Electron emission from projectile ions also contributes to ion-induced electron emission from a solid target. When partially stripped ions are incident on a solid, the accompanying electrons are partly ionized and emitted near the surface, especially when the incident charge is lower than the equilibrium charge [87, 117]. In the most probable case, the electrons are liberated by transfer of the threshold energy for ionization. Therefore, the kinetic energy of the electron in the laboratory frame E_L is approximately equal to that of an electron moving at the velocity of the ion, i.e.

$$E_L = m_e V_1^2/2 = E_B/4 . \tag{6.2}$$

E_L is the so-called *loss-peak energy*, or *cusp energy*. The emission energy of the electrons is sharply distributed around E_L. The shape of the loss peak depends on the bound states from which the electrons are ionized. Fully stripped ions can also produce electrons of kinetic energy $\sim E_L$ by the capture of electrons to the continuum state of the ion [7, 88]. This process is less important for fast, light ions, which have small electron-capture cross sections.

In backward measurements on solid targets, these electrons are emitted from the surface after suffering the scattering processes discussed in Sect. 6.2. Clearly, these electrons contribute to the energy spectrum only in an energy range lower than $\sim E_L$ [67].

7. Auger Electron Emission from Crystals

This chapter is devoted to a review of the studies of Auger electron emission from crystal targets bombarded by fast ions. The periodicity of the crystal lattice affects the electron emission yield through (i) the electron–lattice interaction in the escape process of the electrons, which is a diffraction effect well known in photoelectron spectroscopy, and (ii) the ion–lattice interaction, implying a high-energy shadowing effect in the production process of the electrons. The method of analysis discussed here will be extended to the case of continuum electron yields in later chapters.

7.1 Forward Scattering Effect

The effects of diffraction of ion-induced electrons were investigated in earlier studies of ion–solid interations, as mentioned in Chap. 1. Recently, detailed information about this subject has been obtained by photoelectron spectroscopy. In the following, the diffraction effect of primary interest, namely the forward scattering effect of monoenergetic electrons, is discussed.

7.1.1 Crystal-Assisted Forward Scattering

In Auger or photoelectric processes resulting from inner-shell transitions of atoms, monoenergetic electrons are emitted from lattice sites of the crystal target when it is bombarded by charged particles or photons. Diffraction of these electrons by the atoms in a few surface layers has been widely observed by angle-resolved electron spectroscopy. When the kinetic energy of the electrons is higher than several hundred electron volts, the emission intensity is typically enhanced along the internuclear axes that have short interatomic spacings. This phenomenon is known as the forward scattering effect in Auger electron or photoelectron diffraction. The physical aspects of the forward scattering effect are well understood and are summarized in a review by Egelhoff [89] with emphasis on its application to local-structure analysis of surfaces and interfaces.

Detailed explanation of the forward scattering effect requires a quantum-mechanical treatment since both the wave and the particle nature of electrons appear in this phenomenon [89–91]. Nevertheless, the forward scattering peak can be qualitatively explained by a classical or semiclassical picture, in terms of deflected electron trajectories. In this picture, the peak results from the lens effect of the attractive interaction potential between the electron and the neighboring atom(s), which focuses the originally spherically emitted electron waves towards the internuclear direction, as shown schematically in Fig. 7.1. Rigorously, Auger or photoelectron emission from nonzero-angular-momentum states leads to angular anisotropy in the forward scattering intensity. However, this anisotropy is known to be appreciable only for electrons of kinetic energies below 100 eV.

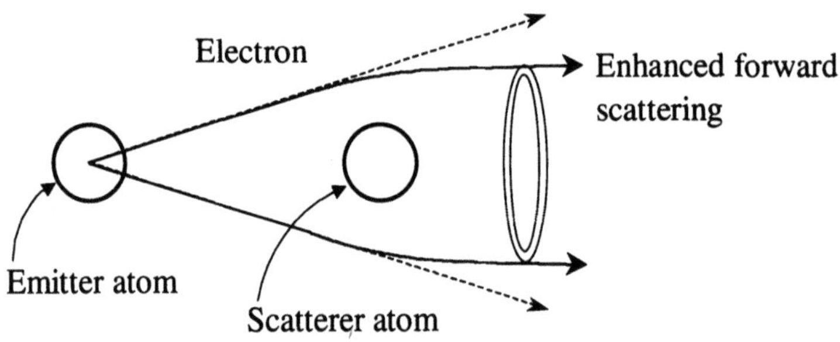

Fig. 7.1. A classical illustration of the forward scattering effect for electrons emitted at energies higher than several hundred eV. This allows a semiclassical explanation of the phenomenon in terms of constructive interference of the scattered electron waves [89]

As another way to understand the forward scattering effect classically, the forward scattering peak can be regarded as a low-energy limit of the electron channeling peak observed for high-energy (typically higher than 100 keV) electrons emitted from radioactive isotopes embedded in substitutional sites of the crystal lattice [92]. This electron channeling corresponds to classical rosette-like trajectories of electrons around an atomic string, which are, essentially, modified focused electron trajectories.

It is important to note that the forward scattering effect is not observed for the continuum electron yield, which originates mainly from electrons that have escaped from a deep region of the crystal, as will be seen later. For these electrons the focused orbits are smeared out after the electrons have suffered elastic and inelastic multiple scattering in the outgoing path. This is contrast to the case of the Auger electron yield, which has escaped through a few surface atomic layers at most. In photoelectron spectroscopy, surface and interface structures can be analyzed only by measuring the direction of the forward scattering peak, which corresponds to the internuclear direction connecting the electron-emitting atom and the scatterer atom(s). In most cases, surface and interface analyses require no complicated calculations including electron multiple scattering, although these are needed to account for the whole angle-resolved pattern including the fine structure in the distribution of the emitted electrons. The forward scattering effect has been observed mainly using energetic photon or electron beams for production of the monoenergetic electrons.

7.1.2 Forward Scattering of Ion-Induced Auger Electrons

We may certainly anticipate that the forward scattering effect will occur for ion-induced Auger electrons. The first observation of this effect for ion-induced Auger electrons was reported by the author's group [93].

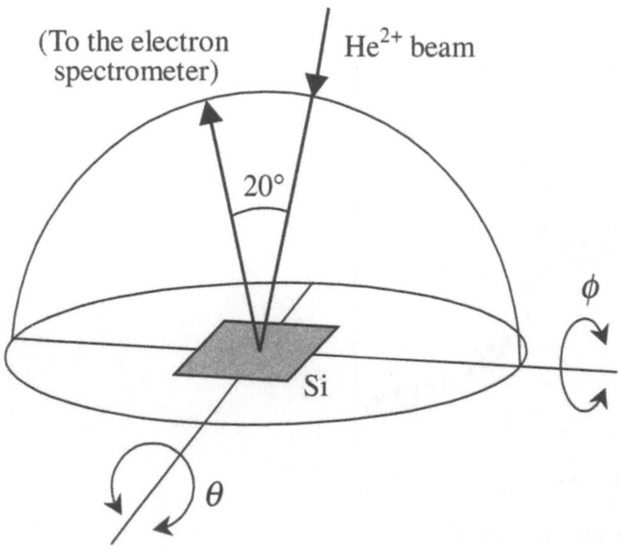

Fig. 7.2. Schematic layout of the measurements of the forward scattering effect [93]

The energy spectrum of Si K-shell Auger electrons has a dominant line at 1.62 keV. This is due to $KL_{2,3}L_{2,3}$ transitions. In the experiments performed by the author's group, Si Auger electrons emitted from hydrogen-terminated surfaces of Si(100) and Si(111) were measured. The ion-induced electrons were energy analyzed at a fixed angle of 160° with respect to the beam direction, using a 45° parallel-plate spectrometer of the double-deflection type described in Sect. 5.1.1. Its angular acceptance of approximately $1° \times 2°$, corresponding to a solid angle of $\sim 6 \times 10^{-3}$ sr, was adequate to obtain a sufficient count rate for the Auger yield. Furthermore, the relative energy resolution of the spectrometer $\Delta E/E \simeq 1.3\%$ resolved the Auger peak well. The target crystals were bombarded by 12 MeV He^{2+} at room temperature under ultrahigh-vacuum conditions (~5×10^{-7} Pa). Prior to the experiments, the axial and planar directions of the Si crystals were determiend from the channeling (high-energy shadowing) dips of the continuum electron yield.

The emission-angle dependence of the Si KLL Auger electrons was measured in terms of the θ and ϕ tilts of the crystal with respect to two orthogonal axes, as shown in Fig. 7.2, using a three-axis goniometer designed for ion channeling experiments.[1] In this case, the rotation angle with respect to the surface normal of the crystal was adjusted so that neither axial nor planar channeling of the incident ions occurred within the 20° × 20° range of θ and ϕ. An angular step of 1° for both θ and ϕ was chosen to observe the enhanced Auger yield expected from typical photoelectron data, which has an angular width of the order of 10°.

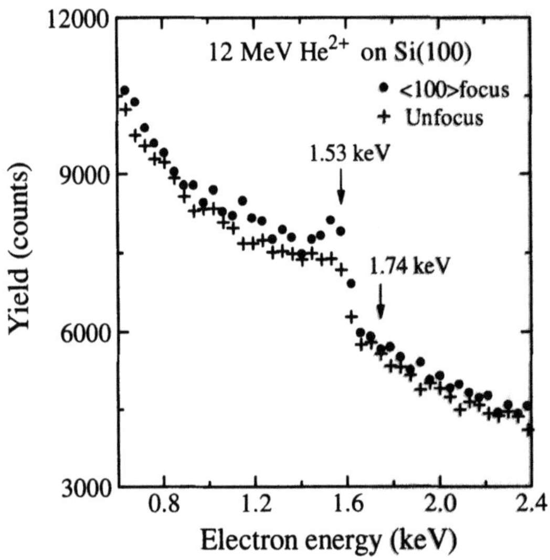

Fig. 7.3. Energy spectra of secondary electrons induced by 12 MeV He^{2+}, measured for Si at two emission angles, corresponding to the ⟨100⟩ focused direction ($\theta = \phi = 0$ in Fig. 7.4) and an unfocused direction. The ions were incident in a random direction. The energy resolution of the spectrometer is ~1.3% [93]

Figure 7.3 shows energy spectra of secondary electrons measured for a Si(100) sample at two emission angles, corresponding to the ⟨100⟩ focused direction (at $\theta = \phi = 0$ in Fig. 7.4) and an unfocused direction (~10° from ⟨100⟩). The Si KLL Auger peak is clearly seen at 1.5–1.6 keV. The angular

---

[1] Tilts θ and ϕ of the spectrometer with a fixed target crystal can provide equivalent data, although angle-resolved measurements of this type tend to suffer from technical difficulties [94].

dependence of the electron yield was measured at 1.53 and 1.74 keV, i.e. at the Auger peak energy and above it.

The Auger electron yield is given by the difference between the electron yields at the two energies. There is a small discrepancy between the yields at the two emission angles at energies far from the Auger peak. This is due to the angular dependence of the counting efficiency for electrons, resulting from instrumental effects. At the 160° observation angle, tilting the crystal caused a slight change in the incidence direction of the electrons into the spectrometer.

Figure 7.4 shows experimental results for the angle-resolved electron yield for Si⟨100⟩. Each pattern of electron yield was constructed from a 20×20 data set. The patterns measured at 1.53 and 1.74 keV are shown in Fig. 7.4a, b, respectively. We see in Fig. 7.4a that the electron yield is enhanced around $\theta = \phi = 0$, where the Si⟨100⟩ axis perpendicular to the crystal surface is aligned with the entrance axis of the spectrometer. However, in Fig. 7.4b no such enhancement is seen except for a slope in the yield due to the instrumental effects noted earlier. The absence of an enhancement in the continuum spectrum has been confirmed for a wide range of experimental parameters [95]. This is consistent with typical observations by photoelectron spectroscopy [96]. For example, Fig. 7.5 shows photoelectron emission spectra from a Ni crystal. In this case, the forward scattering effect can be observed only for those electrons which have escaped elastically or quasi-elastically (for energy losses less than ~20 eV).

Figure 7.4c shows the angle-resolved pattern for the Auger yield, obtained by pixel-by-pixel subtraction of the 1.74 keV yield from the 1.53 keV yield. The pattern for the relative Auger yield, obtained by pixel-by-pixel division of the Auger yield by the 1.74 keV yield, is shown in Fig. 7.4d, in which instrumental effects might be expected to be eliminated.

We see in Fig. 7.4c, d that the forward scattering of the Auger electrons, i.e. an enhanced Auger yield over an angular range of ~10° around the Si⟨100⟩ direction, enhanced by a factor of roughly 1.3 (see also Fig. 7.3). A clear fourfold symmetry, consisting of a square-shaped fringe around the enhanced ⟨100⟩ yield and of (110) ridges, can be recognized in the observed pattern; the directions can be identified from the illustration of the crystallographic directions in the center of Fig. 7.4. Furthermore, a similar pattern of sixfold symmetry has been observed for Si⟨111⟩ [93].

The mean free path for inelastic scattering is ~30 Å for 1.6 keV electrons in Si, according to the calculations by Tung et al. [97]. This value is much longer than the Si⟨100⟩ interatomic spacing (by a factor of ~6). Therefore, the observed patterns should reflect the atomic structure of a thin surface layer, rather than the hydrogen-terminated surface. In addition, the forward scattering effect has not been reported for hydrogen atoms [89], indicating that they hardly deflect Auger electrons or photoelectrons.

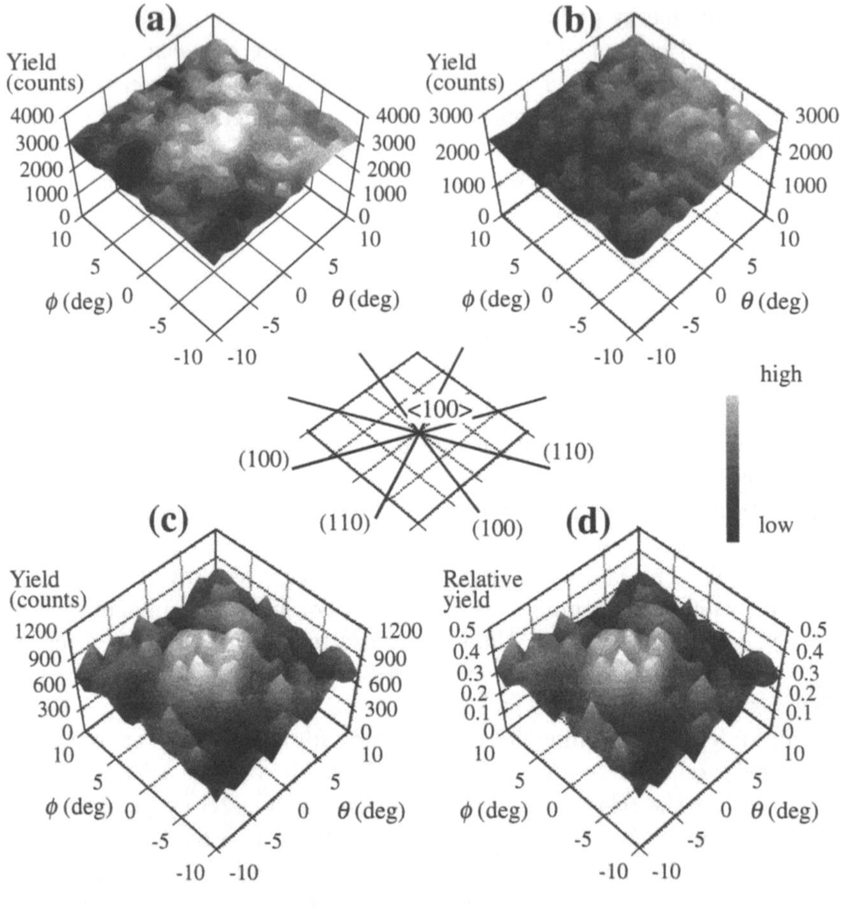

Fig. 7.4. Angle-resolved patterns of secondary electrons induced by 12 MeV He^{2+} for Si(100), observed at energies of 1.53 keV (**a**) and 1.74 keV (**b**). Also shown are the patterns for the Auger yield (**c**) given by the difference between the 1.53 and 1.74 keV yields, and the relative Auger yield (**d**). The *central figure* illustrates the crystallographic directions of the Si sample. For the detailed structure of the patterns, see the original color figures presented in [93]

It is evident that the fourfold ⟨100⟩ fringe of the forward scattering peak is produced not by the atoms along the ⟨100⟩ string containing the electron-emitting atom, but by the atoms in the neighboring strings. Otherwise, the fringe would be of cylindrical symmetry. While the appearance of the crystallographic symmetry outside the forward scattering peak is well known [91,98], the fringe of the peak has not been extensively studied yet. The observed fourfold fringe of the ⟨100⟩ peak can be qualitatively accounted for by modifying the classical picture of the forward focusing of electrons noted earlier. In the

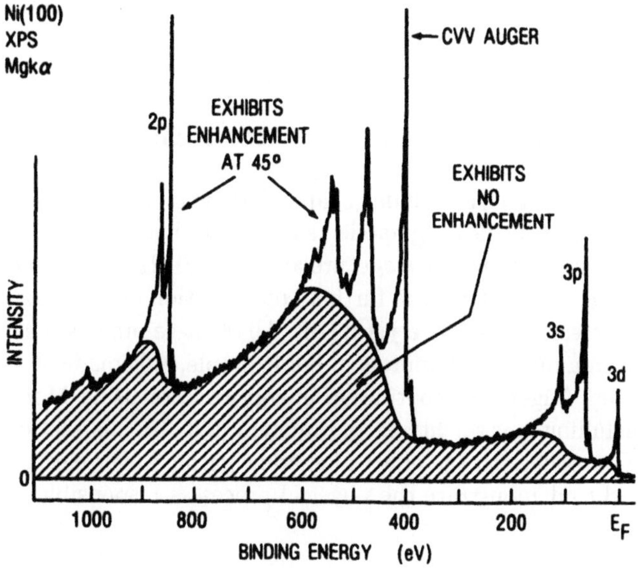

Fig. 7.5. Photoelectron energy spectra emitted from Ni(100), illustrating two components exhibiting and not exhibiting (*hatched*) the forward scattering effect. Note that the electron energy is given in terms of the binding energy (from the Fermi energy E_F) [89, 96]

Si lattice, the ⟨100⟩ atomic strings are more closely spaced in the (110) planes than in the (100) planes by a factor of $\sqrt{2}$. Accordingly, the Auger electrons that move at a small angle from the ⟨100⟩ direction should be more easily caused to diverge from the focused path by the attractive potential of the atoms in the neighboring strings when their orbits are on the (110) planes than when they are on the (100) planes. This should cause a narrower peak width along the directions of the (110) planes than along those of the (100) planes, which is consistent with the observed fringe in Figs. 7.4c,d. More refined experiments are needed to verify the applicability of the classical picture in accounting for the fringe shape.

There are some remarks to be made from the viewpoint of surface analysis. In photoelectron spectroscopy, inner-shell electrons in an energy level of interest can be selectively ionized by monoenergetic photons, while the ionization by fast ions is not level-selective. Nevertheless, there is a unique type of forward-scattering experiment using fast ions in combination with channeling and related effects of the incident ions which allows control of the impact parameters of the ion–atom collisions and, accordingly, localized production of Auger electrons. For example, Auger emission can be localized within the unshadowed surface layer under channeling incidence conditions, as will be discussed in Sect. 7.2, or even in a monatomic layer under grazing-incidence conditions on the crystal surface; see Sect. 11.2.

7.2 Effect of Shadowing on Auger Electron Emission

The ion-induced Auger electron yield under channeling incidence conditions can be accounted for by a shadowing model similar to that used in ion backscattering spectroscopy for surface structure analysis [20–22,66]. Studies of energy-degraded Auger spectra require further consideration of the escape processes of the Auger electrons in the crystal [76]. All of these processes can be taken into account in a numerical approach. For example, Alkemade et al. accounted for measured Auger and continuum energy spectra using Monte Carlo simulations including several fitting parameters [99]. In this section, however, analytical approaches to the effect of shadowing on Auger electron emission are presented to demonstrate the physical processes responsible for the phenomena.

7.2.1 Experimental Conditions

The production of inner-shell Auger electrons from a single-crystal target is affected by the high-energy shadowing of the projectile ions, which reduces close encounters of the ions with target atoms. Because most of the ions (typically more than 95%) are channeled outside the shadowed region surrounding the row of atoms, the inner shells of the crystal atoms are seldom ionized except in the surface layer that is exposed to the ion beam. Under such conditions, the production of ion-induced Auger electrons should be localized near the surface of the crystal. Since this is equivalent to Auger production from a thin target, we may anticipate that the Auger peak will become narrow under channeling incidence conditions.

Alkemade et al. pointed out a possible influence of the forward scattering effect in the observation of ion-induced electrons from crystal targets [99]. Indeed, the observation angle must be chosen when one wishes to investigate the high-energy shadowing effect on the Auger yield without the yield being influenced by the forward scattering effect. For measurements using the commonly used cylindrical-mirror analyzer [20, 21], for example, the angular anisotropy due to the forward scattering effect should be smoothed out by the wide acceptance angle so that the influence of the forward scattering effect can be suppressed. In the opposite case, a spectrometer with a narrow angular acceptance can be used if the channeling and random spectra are measured so that they suffer the same influence of the forward scattering effect. In earlier experiments, the Si KLL Auger yield under ⟨110⟩ channeling incidence conditions was measured at a backward angle of 120° with respect to the beam direction [22]. In this case, no crystal axis coincided with the measured direction, so that there could be no enhanced yield due to the forward scattering effect. In later experiments the electrons were measured at 180° with respect to the beam direction [66,76]. In this layout, any channeling incidence would give rise to an enhanced Auger yield in the direction observed. In the analysis of the data, however, no difficulty occurs if the random yield is measured within ~5° from the channeling direction (see Fig. 7.4), since the physical quantity to be discussed is the ratio of the channeling to the random Auger yield, for which the enhancement factor due to the forward scattering effect can be effectively canceled.

7.2.2 Energy-Degraded Auger Spectra

A fast ion beam generates inner-shell vacancies around its path through a solid. These vacancies are followed by emission of Auger electrons, which is expected from atomic fluorescence-yield curves [100] to be dominant over photon emission when the transition energy is less than ~2 keV. In the escape process from the solid, both elastic and inelastic scattering degrade the initial energy of the Auger electrons and thus need to be considered in accounting for the observed line shape. The principal inelastic processes at these electron energies is plasmon excitation. Elastic scattering, however, can also affect the observed energy spectrum by increasing the total path length of the electron through the solid [101]. To estimate the increase of the total path length due to the cumulative effect of small-angle elastic scattering events, Monte Carlo methods can be used to calculate the ratio between the projected path length $\mathcal{R}_z$ and the actual total path length $\mathcal{R}$. This has been done for 1.5 keV electrons traversing silicon; in the calculation, the elastic scattering cross sections calculated by Riley et al. were used [102].

Figure 7.6 shows the electron distribution with respect to $\mathcal{R}_z/\mathcal{R}$ for $\mathcal{R} =$ 100 and 200 Å. It should be noted that the mean energy loss of 1.5 keV electrons in Si is about 1.6 eV/Å [97], so that the electrons lose ~0.3 keV energy in traversing a path length $\mathcal{R} = 200$ Å. The energy dependence of the elastic scattering cross section in the energy region of interest is too weak to cause any noticeable difference from the results shown in Fig. 7.6 for other energies in this range. We see that $\mathcal{R}_z/\mathcal{R} \geq 0.9$ is satisfied by about 74% of the electrons for $\mathcal{R} = 100$ Å, so that we may neglect the effective increase of the electron path length caused by angular deflection for $\mathcal{R} \lesssim 100$ Å.

Inelastic processes are now considered. We let $P_N(\mathcal{R})$ denote the probability that an electron suffers N inelastic events while traversing the path length $\mathcal{R}$. The probability for $N = 0$ is given by $P_0(\mathcal{R}) = \exp(-\mathcal{R}/\lambda)$, with λ being the mean free path for inelastic scattering. Since inelastic scattering events occur at a rate λ^{-1}, we may write[2]

$$P_N(\mathcal{R}) = \int_0^{\mathcal{R}} \lambda^{-1} P_{N-1}(z) P_0(\mathcal{R} - z)\,\mathrm{d}z\,. \tag{7.1}$$

[2] The integrand of (7.1) implies that the electron suffers its Nth inelastic scattering at a distance $z\ (\leq \mathcal{R})$.

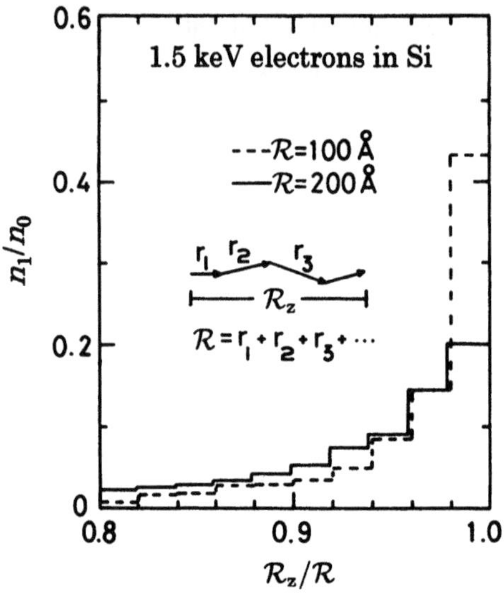

Fig. 7.6. Calculated distribution of the ratio of the projected to the actual path length for 1.5 keV electron in Si. The normalized number of deflected electrons n_1/n_0, with n_0 being the number of electron paths calculated, is shown as a function of $\mathcal{R}_z/\mathcal{R}$ [22]

A mathematical reduction procedure applied to (7.1) leads to the Poisson distribution

$$P_N(\mathcal{R}) = \frac{(\mathcal{R}/\lambda)^N}{N!} \exp(-\mathcal{R}/\lambda) . \tag{7.2}$$

In the above treatment the electron-energy dependence of λ was neglected because the energy loss of the Auger electron under consideration is small compared with the Auger emission energy.

We now consider channeling incidence of the ions. As a result of the ion-beam shadowing effect, the probability of producing a K vacancy in a target atom and hence the probability for Auger electron emission by that atom depends on the distance z from the surface. Let $Q(z)$ be the probability of Auger electron emission at z; this is normalized so that $Q(0) = 1$. The energy spectrum of the observed electrons can be expressed in terms of $Q(z)$ and $P_N(\mathcal{R})$. For Si, plasmon excitation is an important scattering mechanism in the bulk. The energy transferred from an electron is 17 eV for plasmon excitation in Si [103]. For single collisions with valence electrons and for L-shell ionization, the energy transfers are large ($\gtrsim$100 eV). Because we are concerned only with total energy losses less than $\sim$200 eV, the analysis can be simplified by assuming that after N inelastic scattering events, an electron loses an energy NE_{m}, where E_{m} is the mean energy of the electronic excitations produced by the electron in the crystal. For an observation angle of 180° with respect to the beam direction, the observed energy spectrum for the Auger electrons from the crystal can be expressed as

$$Y_{\mathrm{c}}(E_{\mathrm{A}} - NE_{\mathrm{m}}) = \lambda^{-1} \int_0^\infty Q(\mathcal{R}) P_N(\mathcal{R}) \, \mathrm{d}\mathcal{R} \, , \tag{7.3}$$

where E_{A} is the Auger emission energy. The factor of λ^{-1} in (7.3) is included for normalization when (7.3) is extended to the random case. No shadowing effect is anticipated under random incidence conditions, i.e. $Q = 1$ in (7.3) in this case. It follows that the spectrum in the random case becomes

$$Y_{\mathrm{n}}(E_{\mathrm{A}} - NE_{\mathrm{m}}) = 1 \, . \tag{7.4}$$

Clearly, (7.3) and (7.4) for $N = 0$ represent the yields of elastically emitted Auger electrons. These electrons contribute to the Auger peak heights, which are usually detected in an energy-differential mode in the commonly used technique of Auger electron spectroscopy using an energetic electron beam. In this manner, MacDonald et al. successfully interpreted the KLL Auger peak heights observed for Si and Al crystals under channeling incidence conditions using 1–2 MeV H^+ and He^+ [20].

There are some comments to be made on the above model. First, the expression (7.3) can be extended to an observation angle other than 180°, by replacing $\mathcal{R}$ in the Poisson distribution in (7.3) by a modified expression for the outgoing distance [22]. Second, the integration in (7.3) with respect to $\mathcal{R}$ can be replaced by a summation with respect to the discrete intervals $\mathcal{R} = jd$ ($j = 0, 1, 2, \ldots$), where d is the interatomic spacing along the crystal axis. Third, $Q(\mathcal{R})$ is generally obtained by means of numerical calculations of the type described in Sect. 4.6. For a given experimental condition, therefore, the energy-degradad spectra can be calculated as a function of the unknown parameter E_{m} if the parameter λ is known.

7.2.3 Multiple Elastic Scattering

The theoretical considerations in Sect. 7.2.2 can be refined by taking into account a correction factor for the multiple elastic scattering of the Auger electrons in the crystal [76]. When an Auger electron emitted at a depth $\mathcal{R}$ reaches the surface along an outgoing path that is not straight owing to multiple scattering, the effective path length of the electron $\mathcal{R}_{\mathrm{e}}$ is distributed over values of $\mathcal{R}_{\mathrm{e}} > \mathcal{R}$. We can estimate an analytical expression for the average value of the enhanced path length $\langle \mathcal{R}_{\mathrm{e}} \rangle$ as a function of $\mathcal{R}$. For a monoenergetic electron emitted in a homogeneous medium, the time evolution of various spatial averages can be evaluated from transport equations [101,104]. At time t the mean penetration distance $\mathcal{R}_{\mathrm{p}}$ of an electron emitted at $t = 0$ with velocity v_0 is approximately given by

$$\mathcal{R}_{\mathrm{p}} = \lambda_{\mathrm{t}} \left[1 - \exp(-v_0 t / \lambda_{\mathrm{t}}) \right] \, , \tag{7.5}$$

where λ_t is the transport mean free path of the electron, given by

$$\frac{1}{\lambda_t} = n_a \int_0^{\pi} (1 - \cos\phi_s) d\sigma(\phi_s) . \tag{7.6}$$

Here n_a is the number density of atoms in the crystal, and $\sigma(\phi_s)$ is the differential cross section for elastic scattering at a scattering angle ϕ_s. Since $v_0 t = \mathcal{R}$ for a straight path, we may anticipate, for a small enhancement of the outgoing path length, that

$$\mathcal{R}/\mathcal{R}_p \simeq \langle \mathcal{R}_e \rangle / \mathcal{R} , \tag{7.7}$$

i.e.

$$\langle \mathcal{R}_e \rangle = \frac{\mathcal{R}^2}{\lambda_t \left[1 - \exp(-\mathcal{R}/\lambda_t)\right]} . \tag{7.8}$$

It follows that a modified alternative to (7.3) can be written

$$Y_c(E_A - N E_m) = \lambda^{-1} \int_0^{\infty} Q(\mathcal{R}) P_N(\langle \mathcal{R}_e \rangle) \, d\mathcal{R} , \tag{7.9}$$

For $\lambda_t = \infty$, i.e. when multiple angular deflection can be ignored, (7.9) coincides with (7.3) since $\langle \mathcal{R}_e \rangle$ becomes equal to $\mathcal{R}$.

7.2.4 Determination of the Electron Stopping Power

From a comparison of the calculated and observed degraded Auger spectra, the parameter E_m and, therefore, the electron stopping power for the Auger electrons S_e can be determined from the relation

$$S_e = E_m / \lambda . \tag{7.10}$$

This provides a unique technique for measurement of the stopping power of low-energy electrons (typically lower than 2 keV), distinct from foil and thin-layer transmission experiments [105, 106].

The analysis procedure has been demonstrated for Si KLL Auger electrons [22, 66]. Figure 7.7 shows the secondary-electron energy spectra for 18 and 24 MeV He^{2+} incident in the $\langle 110 \rangle$ axial direction and in a random direction of Si, measured for the same number of incident He^{2+}. The energy-degraded spectra for the channeling case were obtained by assuming the baselines shown by the dashed curves. In general, the background subtraction must be carried out carefully for a random case or a noncrystalline target, taking into account the long low-energy tail of the energy-degraded Auger yield [107]. For the channeling case, however, the background yield is relatively easy to estimate since the shadowing effect causes a much shorter low-energy tail than for the random case.

Each pair of vertical arrows in Fig. 7.7 indicates the yield of elastically scattered Auger electrons for the random case. The ratio of the yields for 24

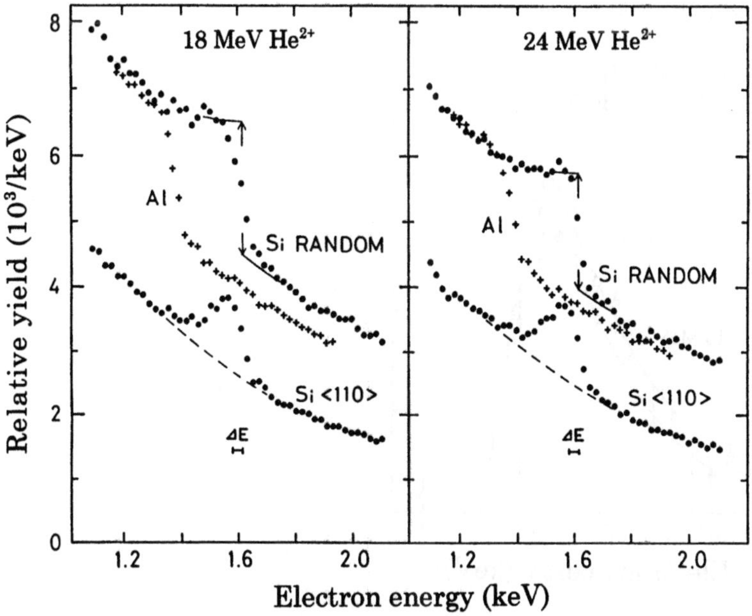

Fig. 7.7. Energy spectra of secondary electrons induced by 18 and 24 MeV He^{2+} incident in the ⟨110⟩ direction and in a random direction of Si, measured with the experimental setup shown in Fig. 5.1. The *dashed curves* show the assumed baselines of the Si KLL Auger spectra [66]

and 18 MeV He^{2+} is 0.90 ± 0.03. This value is in agreement with the corresponding ratio, 0.94, of the Si K-shell ionization cross sections [108]. The long low-energy tails of the Auger electron yield seen for the random case result essentially from multiple elastic scattering, which reduces more effectively the number of energy-degraded Auger electrons escaping from deeper regions. This decrease may, however, be neglected for the present energy range of interest, i.e. 1.4–1.6 keV; see also Fig. 7.6. Another indication of the validity of this assumption is given by the spectra for an Al polycrystalline target, also shown in Fig. 7.7. The baselines of the Si Auger yield should be approximately parallel to the continuum yield for Al at energies higher than ~1.4 keV, which is the dominant line of Al K-shell Auger electrons.

Figure 7.8 shows normalized spectra of energy-degraded Si KLL Auger electrons for Si⟨110⟩, obtained by dividing the Auger yield by the elastically emitted Auger yield in the random case. For such a narrow energy range (1.4–1.6 keV), the detection efficiency of the electron multiplier, referred to in Sect. 5.1.3, can be assumed to be constant. The calculated spectra as a function of N were obtained from (7.3) by adopting the theoretical value of $\lambda = 32$ Å for 1.6 keV electrons in Si [97, 103]. The calculated results for $Q(\mathcal{R})$ in the present case have been previously shown in Fig. 4.10. The value of $E_{\rm m}$ was chosen so that the calculated spectrum, in which the yield was given

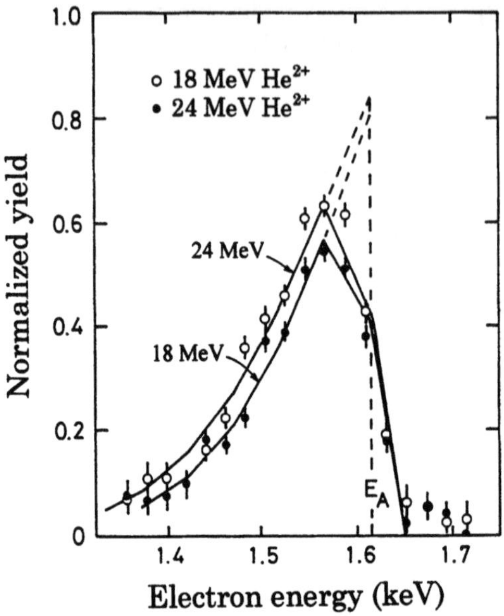

Fig. 7.8. Comparison between measured (*open* and *filled* circles) and calculated (*solid lines*) spectra of the energy-degraded Si KLL Auger electrons for ⟨110⟩ incidence on Si. E_A (= 1.62 keV) is the dominant line of the Si K-shell Auger emission. The calculated spectra, fitted for E_m = 49.1 eV, include a correction for the spectrometer's finite energy resolution. The *dashed lines* show the uncorrected spectra [66]

at an energy interval of E_m, reproduced the observed spectrum shape. The calculated spectra for the fitted value $E_m = 49.1$ eV are shown in Fig. 7.8 by the solid lines, which include a correction for the finite energy resolution of the spectrometer. The correction is appreciable only for $N = 0$, i.e. at E_A. The value of E_m finally obtained was 48.0 ± 2.7 eV, from which the stopping power was determined as $S_e = 1.50 \pm 0.08$ eV/Å using (7.10). This is in good agreement with the theoretical value of $S_e = 1.6$ eV/Å given by Tung et al. [97]. It is noteworthy that in the present case the Bethe–Bloch formula for the stopping power gives a reasonable value of $S_e \simeq 1.4$ eV/Å,[3] although the applicability of the formula to such low-energy electrons is not clear [24,109]. The above results are consistent with results previously obtained from similar analyses of the Auger electron yield observed at a backward angle of 120° [22].

It should be emphasized that the determination of the electron stopping power described here is insensitive to the choice of λ. This can be qualitatively explained by the fact that the average number of inelastic scatterings N_{av} suffered by the electron when traversing a thin layer is proportional to λ^{-1}.

[3] A value of 159 eV for the mean ionization potential of Si was used [109]. The shell correction was not included.

For the fitted value of E_m, the quantity $N_{av}E_m$ should be roughly equal to the observed energy width of the Auger spectrum in the channeling case. It follows that E_m/λ, i.e. the stopping power obtained, is independent of the initial choice of λ in the analysis procedure.

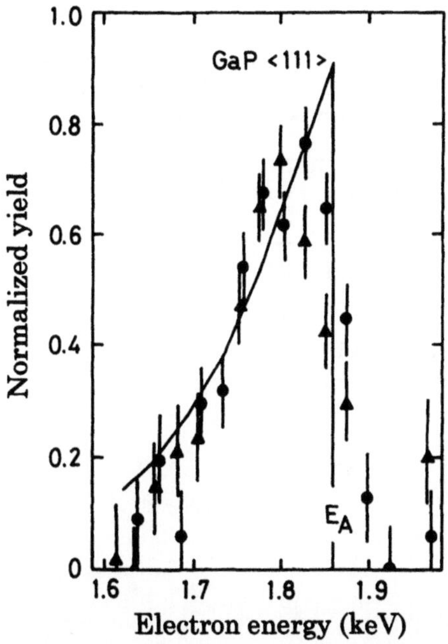

Fig. 7.9. Comparison between measured (*filled circles* and *triangles*) and calculated (*solid line*) spectra of energy-degraded P KLL Auger electrons for ⟨111⟩ incidence on GaP. E_A (= 1.86 keV) is the dominant line of the P K-shell Auger emission. The parameters assumed in the calculated spectrum were E_m = 41 eV, λ = 25 Å, and λ_t = 220 Å [76]

Similar measurements and analyses have been reported for Mg KLL Auger electrons (E_A = 1.18 keV) in MgO crystals, and for P KLL Auger electrons (E_A = 1.86 keV) in GaP crystals [76]. Figure 7.9 compares the observed P KLL Auger yield with the yield calculated from (7.9) under GaP⟨111⟩ channeling incidence conditions for 20 MeV He^{2+}. The correction for multiple angular deflection typically causes a decrease in the determined value of S_e relative to the uncorrected case. The factors of decrease are typically 0.85 and 0.95 for analyses of Mg and Ga KLL Auger electrons, respectively. The corresponding factor for Si has a relatively large value of ~0.96, as in the case of GaP, which is due to the large value of λ_t = 330 Å [102] compared with $\lambda \simeq 30$ Å. In such cases, the angular deflection has a minor influence on the determination of the electron stopping power.

The values of S_e obtained from (7.9) assuming theoretical values of λ_t are shown in Table 7.1, together with the earlier results for Si. We may conclude that the stopping powers obtained can be reasonably accounted for in terms of a low-energy extrapolation from those estimated from the Bethe–Bloch stopping formula.

Table 7.1. Electron stopping powers S_e determined from Auger electron spectroscopy under channeling incidence conditions. The calculated values S_{cal} were obtained from the Bethe–Bloch formula, unless otherwise noted [22, 66, 76]

Electron energy (keV)	Crystal (stopping medium)	S_e (eV/Å)	S_{cal} (eV/Å)
1.18	MgO	2.10 ± 0.21	2.7
1.62	Si	1.50 ± 0.08	1.4, 1.6[a]
1.86	GaP	1.60 ± 0.15	1.7

[a]Tung et al. [97]

8. Binary-Encounter Electron Emission from Crystals

This chapter deals with the continuum energy spectra of electrons emitted from crystal targets bombarded by fast ions under channeling and random incidence conditions. The theoretical models used for understanding the phenomena are, in part, an extension of those introduced for the analysis of monoenergetic electron emission in Chap. 7. Note that the experimental electron spectra presented here are raw data, i.e. the electron yield is the number of electron signals counted, as noted in Sect. 5.1.1.

8.1 Binary-Encounter Yield

Experimental evidence for the binary-encounter production of ion-induced electrons in solids can be easily recognized in a comparison of continuum energy spectra measured for equal-velocity ions. Figure 8.1 shows energy spectra of the secondary electrons induced by 3.75 MeV/u $^{2}H^{+}$ (deuterons), $^{4}He^{2+}$, $^{28}Si^{13+}$, and $^{35}Cl^{11+}$ for random incidence on a Si crystal. The electron yields are normalized to the same number of incident ions and also to the square of the atomic number of the ion Z_1. In this case, the binary-encounter peak energy is 8.2 keV, so that the measured spectra cover most of the energy range to which both inner-shell and valence electrons contribute. Random spectra seldom depend on the input charge state of the incident ions except at energies lower than $\sim E_L$, equal to 2 keV in the present case, since the binary-encounter electrons are effectively induced by the ions in their equilibrium charge states (Sect. 9.5).

We see that the spectra shown in Fig. 8.1 are identical. They demonstrate the Z_1^2 scaling of the electron emission from solid targets at energies above ~2 keV, within an estimated uncertainty of 20% in the vertical scale, despite the wide range of variation of Z_1^2 from 1 (for $^{2}H^{+}$) to 289 (for $^{35}Cl^{11+}$). The 20% discrepancy may stem from both experimental error and the screening effect of the partially stripped heavy ions (see also Sect. 9.5). For the screened case, the scaling parameter Z_1 should be replaced by a modified parameter. Apart from the relative uncertainty in the vertical scale, there is no discernible difference between the shapes of the spectra shown in Fig. 8.1 at energies higher than ~2 keV, as can be seen by overlapping them by a vertical shift.

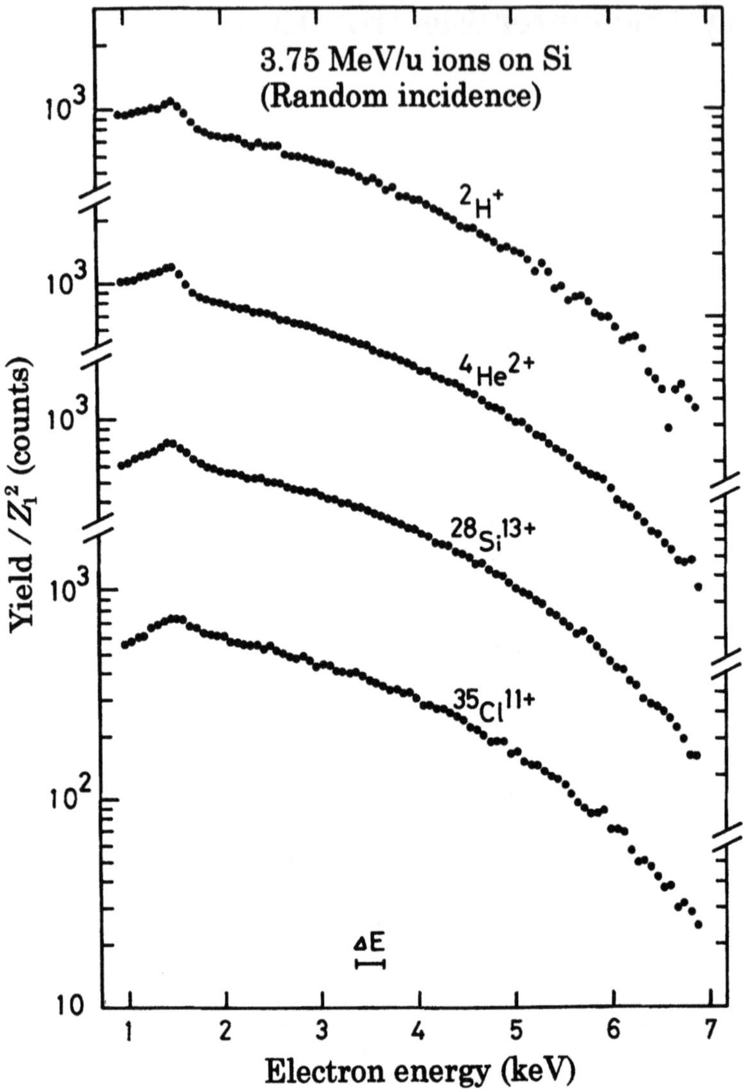

Fig. 8.1. Energy spectra of secondary electrons induced by 3.75 MeV/u deuterons and He, Si, and Cl ions, measured for random incidence on a Si crystal. The peaks near 1.6 keV are due to Si KLL Auger electrons [86]

Similar results have also been obtained for lower-velocity ions. Figure 8.2 shows energy spectra of electrons induced by 92.5 keV/u H^+ and He^+, measured again for Si under random incidence conditions. The vertical scale is normalized similarly to the case of 3.75 MeV/u ions shown in Fig. 8.1. The spectra coincide each other, demonstrating the Z_1^2 scaling of the energy spectra. According to a semiempirical formula used for estimating ion stopping

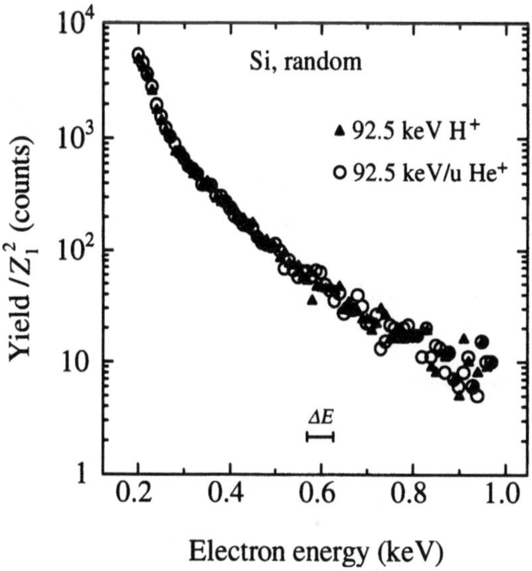

Fig. 8.2. Energy spectra of secondary electrons induced by 92.5 keV/u H^+ and He^+, measured under random incidence conditions [85]

powers [110, 111], 92.5 keV/u He^+ ions in solids are not fully stripped, but have a reduced effective charge of ~1.6. The observed Z_1^2 scaling indicates that for such light ions the captured inner-shell electron(s) hardly affects the electron emission, and therefore the ions can be effectively assumed to be fully stripped in the electron production process.[1]

The interpretation of the observed Z_1^2 scaling is straightforward. For fully stripped ions of equal velocity, the double differential cross section (with respect to energy and angle) for recoil of atomic electrons is proportional to Z_1^2, as concluded in Sect. 3.4. The energetic electrons produced generally experience elastic and inelastic scattering before they are emitted from the surface, by which process energetic electrons can also be produced. For equal-velocity ions, all of the processes, including production and escape of the electrons, are identical except for the scaling factor of Z_1^2, which merely increases the number of electrons participating in the processes. We may therefore expect that the Z_1^2 scaling for binary-encounter production of the electrons is preserved in the electron energy spectra measured with solid targets.

---

[1] As will be discussed in Chap. 9, enhanced electron emission by partially stripped ions is anticipated. However, this effect is of practical importance only for heavy ions.

8.2 Normalized Channeling Yield

The high-energy shadowing effect influences the processes of production of binary-encounter electrons in a manner that is essentially similar to that for Auger electrons, as discussed in Sect. 7.2. The unshadowed surface layer is a source of binary-encounter electrons [67, 86, 99], while the inner shells of the target atoms in deep regions of the crystal are effectively shadowed. As a result, the production of binary-encounter electrons is suppressed under channeling incidence conditions compared with the random case.

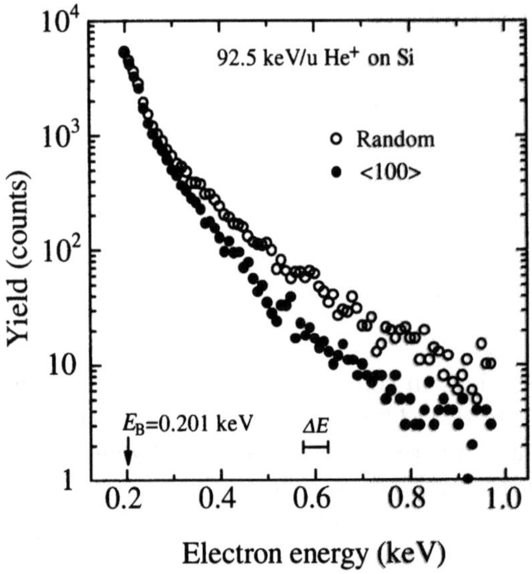

Fig. 8.3. Energy spectra of secondary electrons induced by 92.5 keV/u He^+ for Si under ⟨100⟩ and random incidence conditions [85]

It is of general interest to survey the experimental results obtained for ion energies ranging from 0.1 to 10 MeV/u. Figure 8.3 shows the energy spectra of electrons induced by 92.5 keV/u He^+, measured under Si⟨100⟩ and random incidence conditions [85]. The yield in the channeling case is clearly reduced at energies higher than ~0.3 keV. This is caused by the high-energy shadowing effect mentioned earlier. It should be pointed out that the electrons lost from He^+ contribute to the spectrum only at energies lower than $\sim E_L (= 0.05\,\text{keV})$, as noted in Sect. 6.3. Figure 8.4 shows similar energy spectra induced by 3.5 MeV/u O^{8+} [85]. Even at energies lower than $E_B = 7.6\,\text{keV}$, the channeling yield is appreciably lower than the random yield, in contrast to the case of 92.5 keV/u He^+, for which the shadowing effect can hardly be recognized below $E_B = 0.201\,\text{keV}$. The experimental results for higher-velocity ions are essentially similar, for example, for 6 and 8 MeV/u H^+ and He^{2+} [44, 84].

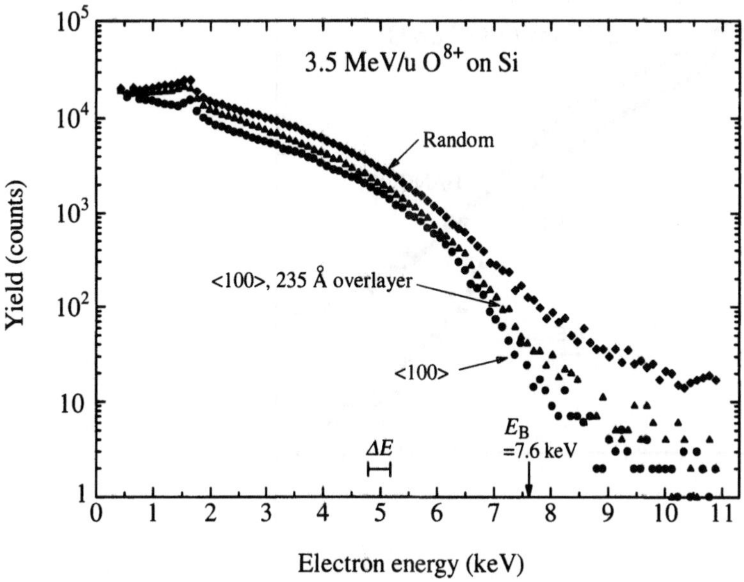

Fig. 8.4. Energy spectra of secondary electrons induced by 3.5 MeV/u O^{8+} for Si with and without a noncrystalline Si overlayer 235 Å thick under Si⟨100⟩ channeling and random incidence conditions [85]

A normalized spectrum can be obtained by dividing the channeling yield by the random yield at the same electron energy, as demonstrated in Fig. 8.5 for the energy spectra of electrons emitted from a Ge crystal bombarded by 2.5 MeV/u O^{5+}. In this procedure the energy dependence of both the acceptance of the spectrometer and the electron detection efficiency, discussed in Chap. 5.1, are eliminated. Moreover, the effective cross section for electron emission from the surface, which varies with energy and angle, is also canceled, so that the absolute value of the cross section can remain unknown in the analysis of the normalized yield.

Generally, a remarkable decrease in the normalized yield is seen at energies higher than $\sim E_B$ (Fig. 8.5). Further examples will be found in the experimental data shown in later chapters, e.g. in Figs. 9.3 and 10.4. This decrease is due to the strong effect of shadowing on the inner-shell electrons [67]. At energies lower than E_B the unshadowed valence or outer-shell electrons enhance the normalized yield. Such a step structure at $\sim E_B$ is consistent with the backscattered energy spectra of energetic electrons from solid targets, which result predominantly from elastic and quasi-elastic scattering processes, as mentioned earlier in Sect. 6.1.

At a fixed electron energy the normalized channeling yield, i.e. the ratio of the channeling to the random yield W, can be used as a measure of the high-energy shadowing effect. Typically, the channeling yield is enhanced by the

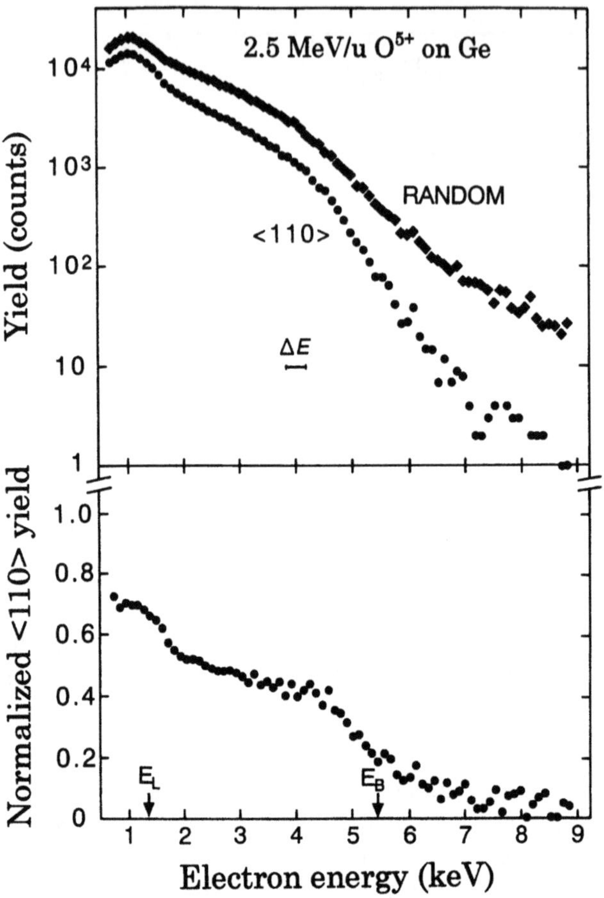

Fig. 8.5. Energy spectra of secondary electrons from Ge induced by 2.5 MeV/u O^{5+} under ⟨110⟩ channeling and random incidence conditions. The *lower part* shows the normalized ⟨110⟩ yield [69]

angular divergence of the ion beam, so that care must be taken to determine the exact value of W for a parallel beam. The determination of W for zero angular divergence of the ion beam requires a beam collimation technique using a four-blade collimator. Figure 8.6 shows an example of measurements of W at energies above and below $E_B = 5.45$ keV for a Ge crystal. We see that the value of W decreases with decreasing beam current, i.e. as the collimator is narrowed. The slopes for the 3.3 keV yield are greater than those for the 7.7 keV yield, indicating that the shadowing for all electrons is more sensitive to the angular divergence of the beam than that for the inner-shell electrons. This is also consistent with observations of angular dips for GaP and InP at energies above and below E_B, as will be shown in Figs. 10.5 and 10.6 in Sect. 10.3. The wider critical angle for the inner-shell shadowing corresponds

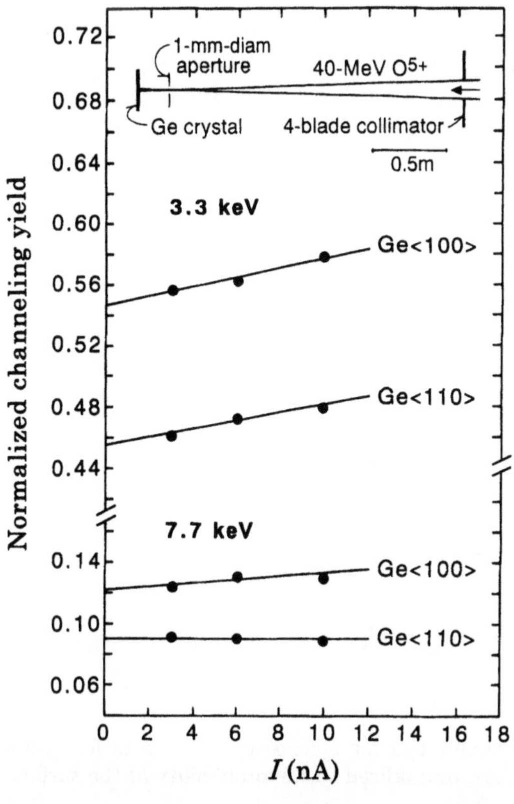

Fig. 8.6. Dependence of the normalized channeling yield on the beam current I on the target, for Ge. I was reduced by narrowing the four-blade collimator shown in the *inset*. Note that the slopes of the lines, drawn to guide the eye, should depend on the angular divergence of the beam. The normalized channeling yield for zero angular divergence can be obtained by extrapolation to $I = 0$ [69]

to the critical angle for channeling commonly observed using backscattered ions. The exact value of W for zero angular divergence of the beam can be obtained from Fig. 8.6 by extrapolating the measured value of W to zero beam current ($I = 0$).

8.3 Two-Component Model of the Electron Yield

Valence and loosely bound electrons are distributed rather uniformly in a crystal lattice, and therefore, even in the channeling case, the binary-encounter events for these electrons should occur at a constant rate with respect to the depth from the surface. Accordingly, we may assume two components for the electron yield: one originates from inner-shell electrons that are well shadowed, while the other originates from the unshadowed electrons referred to above. The unshadowed electrons contribute to the spectrum in an energy range lower than E_{B}. In the following discussion, therefore, the normalized electron yield in an energy range lower than E_{B} but, of course, higher than $E_{\mathrm{L}} = E_{\mathrm{B}}/4$ is of interest.

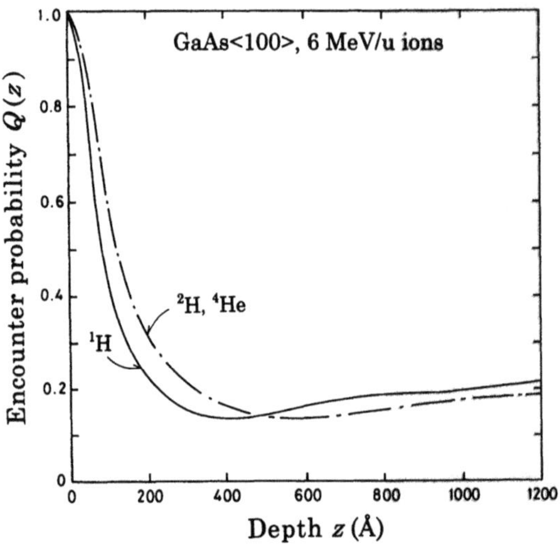

Fig. 8.7. Calculated encounter probabilities for all electrons in GaAs for ⟨100⟩ incidence of 6 MeV/u ^{1}H, ^{2}H, and ^{4}He, normalized to the probability at the surface [44]

The component of the electron yield that originates from the unshadowed electrons can be estimated experimentally [44, 84]. The principle of the method can be well understood using the numerical simulations discussed in Sect. 4.6, although such simulations include less realistic assumptions, for example uncorrelated thermal displacements of crystal atoms (Sect. 4.6). The spatial electron distribution used in the simulations presented here was obtained by applying the Poisson equation to the Molière potential. Figure 8.7 shows the encounter probabilities for all electrons in GaAs, calculated for ⟨100⟩ incidence of fully stripped 6 MeV/u ^{1}H and ^{2}H (or ^{4}He) at room temperature. Note that the encounter probabilities for equal-velocity ^{2}H and ^{4}He are the same because the value of R_{c} is the same; see Sect. 4.4. The calculated results satisfy the trajectory scaling predicted by the continuum model of channeling, discussed in Sect. 4.5.1. We see that the probability curve for protons coincides with that for deuterons or ^{4}He if it is enlarged with respect to the depth by a factor of 1.4. This value corresponds to the ratio of the trajectory scaling parameter $V_1\sqrt{M_1/Z_1}$ for ^{2}H (or ^{4}He) to that for ^{1}H, i.e. $\sqrt{2} \simeq 1.4$. At a depth $z \gtrsim 500$ Å the encounter probabilities are less dependent on the depth; here the encounter probability corresponds essentially to the contribution from unshadowed electrons such as valence and loosely bound electrons mentioned earlier.

These results clearly demonstrate the validity of the two-component model of the electron yield. The component originating mainly from the unshadowed surface layer for inner-shell electrons is expected to reflect the character of the trajectory scaling. Indeed, it should be proportional to $\sqrt{M_1/Z_1}$ unless the electrons have relatively short escape distances (see Sect. 8.4 for details). The other component μ (< 1), produced from the valence and loosely bound electrons, must be independent of the trajectory scaling parameter. Furthermore, for equal-velocity ions the random spectra are identical except for a factor of Z_1^2 when they are fully stripped in the crystal, according to Sect. 8.1. This allows a comparison of the W values measured for equal-velocity ions at the same electron energy. We may therefore write

$$\frac{W_\mathrm{f} - \mu}{W_\mathrm{p} - \mu} = \sqrt{2}\,, \tag{8.1}$$

where W_f and W_p are the normalized electron yields for fully stripped ^{4}He and protons, respectively, of equal velocity. The factor $\sqrt{2}$ originates again from the trajectory scaling parameter noted earlier. W_f can be also obtained from other equal-velocity ions of $M_1/Z_1 = 2\,\mathrm{u}$, for example ^{16}O, if they are fully stripped in the crystal. The value of μ can be determined from (8.1) by employing the measured values of W_f and W_p.

As will be seen later, the trajectory scaling and the Z_1^2 scaling for equal-velocity ions are key factors in the analysis of channeling spectra by the two-component model. A successful application of the two-component model will be presented in Chap. 9.

8.4 Electron Emission from Overlaid Crystals

The production and escape processes of binary-encounter electrons can be studied experimentally by using a crystal covered with a thin noncrystalline layer which enhances the number of unshadowed atoms near the surface, as schematically shown in Fig. 8.8. By changing the overlayer thickness, the effective production of binary-encounter electrons can be artificially controlled. Measurements and analysis performed with overlaid crystals provide an intuitive understanding of electron emission under channeling and random incidence conditions, which can also be extended to noncrystalline targets.

8.4.1 Thin Unshadowed Layers

An amorphous Si overlayer on a Si crystal can be prepared by deposition of evaporated Si at room temperature under ultrahigh-vacuum conditions (10^{-8}–10^{-9} Pa). By using a typical quartz oscillator thickness gauge, the

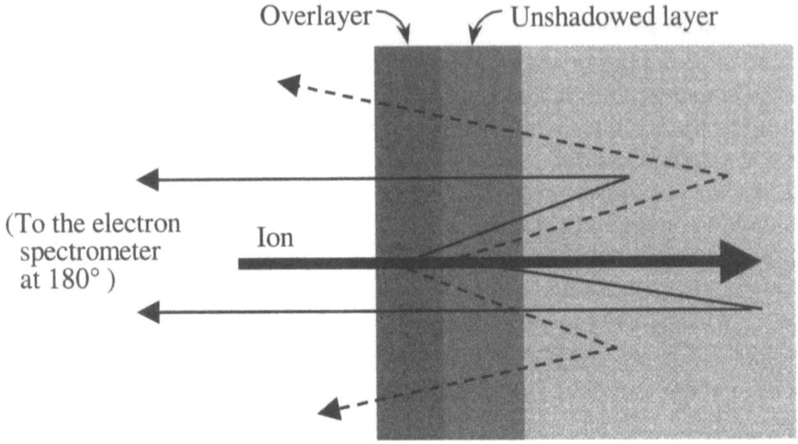

Fig. 8.8. Schematic illustration of binary-encounter electron emission from an overlaid crystal. High-energy electrons are produced by hard recoil (in the forward direction) from the noncrystalline overlayer and the unshadowed crystal layer. They are then backscattered by the crystal atoms and emitted from the surface towards the electron spectrometer. Trajectories of detected and undetected electrons are shown by the *solid* and *dashed lines*, respectively [84]

thickness of the overlayer can be controlled within an uncertainty of ±5% for any given value of the atom density of the overlayer.

The energy spectra of electrons induced by 92.5 keV/u He^+ at a backward angle of 180° were measured for overlaid Si crystals using the experimental apparatus described in Sect. 5.2. The values of W were determined in two energy ranges, of 0.36–0.40 and 0.72–0.80 keV, using the energy spectra shown in Fig. 8.3. Figure 8.9 shows W for the ⟨100⟩ direction plotted against the Si overlayer thickness x. The overlayer thickness measured using a quartz oscillator is given as a mass per unit area, while it is more convenient to express it in Å, for example, for comparison with the effective thickness of the unshadowed crystal layer in the later discussion. Therefore, values of the overlayer thickness obtained by assuming the atom density of crystalline Si, rather than the actual density, are presented in Fig. 8.9.[2]

In Fig. 8.9, the value of W for the 0.72–0.80 keV yield increases linearly up to $W \simeq 0.7$ with increasing overlayer thickness. Such a linear increase is expected to occur when the effective thickness of the unshadowed layer where the forward recoil occurs is given by the sum of the actual thickness of the overlayer and that of the unshadowed crystal layer. On the other hand,

[2] The ratio of the atom density in the evaporated Si used here to that of crystalline Si is estimated to be 0.94–0.98 for the present thickness range of the evaporated layer, according to Brodsky et al. [112].

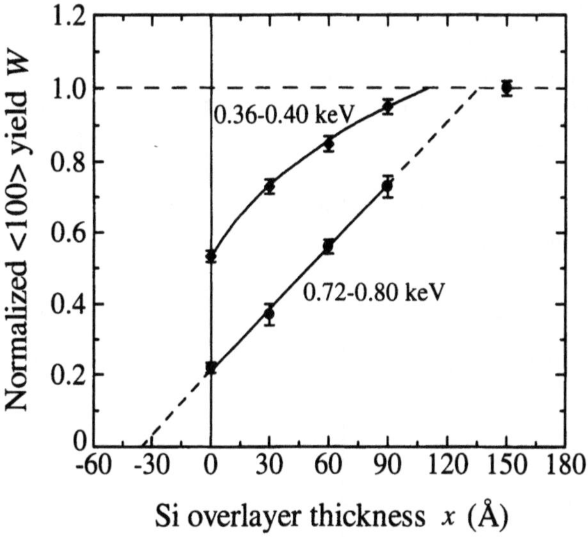

Fig. 8.9. Dependence of W on the thickness of the noncrystalline Si overlayer x, measured with 92.5 keV/u He^+. The *solid* and *dashed lines* are drawn to guide the eye [85]

the 0.36–0.40 keV yield increases less sensitively with increasing overlayer thickness and approaches the random level ($W = 1$).

Figure 8.9 indicates that an overlayer of 135 Å thickness would enhance the value of W for the 0.72–0.80 keV yield to unity, i.e. the random level, as shown by the dashed line. Furthermore, the value of $-x = 35 \pm 5$ Å obtained by extrapolation to $W = 0$, also shown by the dashed line, is the effective surface layer thickness t, i.e. the effective thickness of the surface layer from which the observed electrons originate. Therefore, the effective escape length for the electron yield, D, introduced in Sect. 6.2, can be estimated as the sum of 135 Å and the intrinsic layer thickness $t = 35$ Å, i.e. $D = 170$ Å. Obviously, for $x = 0$ W can be written

$$W = t/D \,, \tag{8.2}$$

which is the intrinsic value of W for the crystal without an overlayer. The value of D determined by this method is characteristic of the electron yield at any fixed energy of interest, irrespective of the channeling incidence condition chosen for its determination. Also, this value is well defined only for 180° measurements of the electron yield because in this case it is independent of the incidence direction of the ion beam on the crystal surface, as was discussed in Sect. 6.2. It is important to note that for oblique incidence of the ion beam on the surface, the effective escape length and, of course, the effective surface layer thickness are measured along the beam axis.

The electrons produced in the crystal lose kinetic energy on their outgoing paths. The maximum energy of electrons produced that contribute to the 0.80 keV yield can be roughly estimated as $0.80 + DS = 1.14$ keV, where $S \approx 2$ eV/Å is the stopping power of Si for 0.8–1.1 keV electrons [97]. It must be noted that in this estimate a straight outgoing path of the electron is assumed; this is based on the discussion in Sect. 6.2. In a similar manner, a maximum energy of ~0.6 keV is obtained for the 0.4 keV yield ($S \approx 3$ eV/Å and $D = 66$ Å, see later discussion).

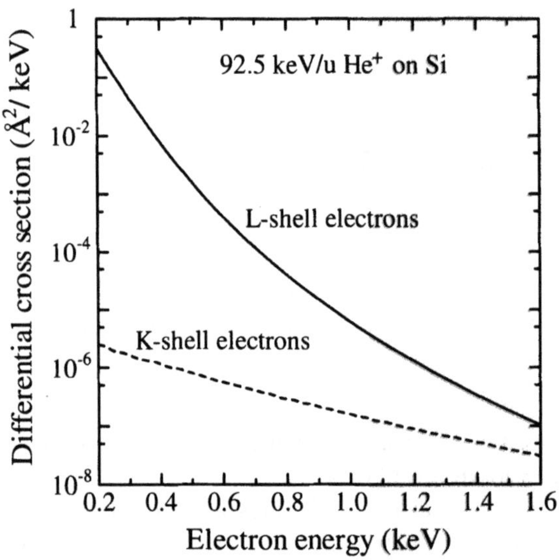

Fig. 8.10. Differential cross sections $\mathcal{Y}_2(E_e)$ for recoil of K- and L-shell electrons from a Si atom by impact of 92.5 keV/u He^+, calculated from (3.22). Note that the number of bound electrons in each shell has been taken into account [85]

For further consideration of the results shown in Fig. 8.9, the production of the binary-encounter electrons has been investigated theoretically. Figure 8.10 shows the calculated differential cross sections, i.e. the energy spectrum $\mathcal{Y}_2(E_e)$ given by (3.22), for recoil of Si K- and L-shell electrons produced by 92.5 keV/u He^+. In this calculation, the binding energies for Si were taken to be 1839, 149, and 99 eV for the K, L_1, and $L_{2,3}$ levels, respectively [113]. The calculated results shown in Fig. 8.10 indicate that the observed electrons are mainly produced from Si L shells. In an energy range of interest of 0.8–1.14 keV, for example, the ratio of the integrated differential cross sections for the K and L shells amounts to only 0.013, implying that the K-shell electrons contribute negligibly to the observed yield. Consequently, the W versus x relations shown in Fig. 8.9 can be attributed to the effect of high-energy shadowing on Si L-shell electrons.

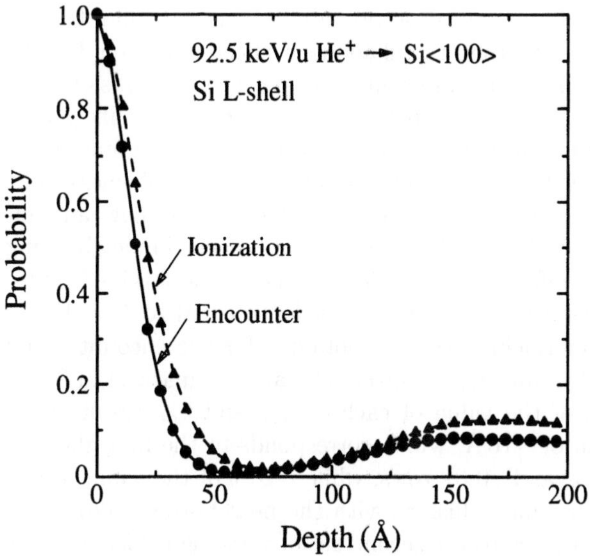

Fig. 8.11. Calculated encounter probability and ionization probability for Si L-shell electrons under Si⟨100⟩ channeling incidence conditions for 92.5 keV/u He^+, normalized to the corresponding probabilities for a surface atom. The probabilities, calculated at intervals of 5.43 Å, the ⟨100⟩ interatomic spacing, are shown by the *solid* and *dashed curves* to guide the eye [85]

For precise analysis of the experimental results, the impact-parameter dependence of the probability for hard recoil of Si L-shell electrons must be obtained theoretically. In the present case, however, a different approach can be used to verify the L-shell shadowing. The value of $t = 35 \pm 5$ Å for the 0.72–0.80 keV yield can be accounted for by using the calculated depth dependence of the Si L-shell encounter probability and ionization probability under channeling incidence conditions. It is of further interest to investigate which probability is better for describing the recoil of the L-shell electrons that contribute to the observed electron yield. Details of the numerical simulations of channeling, based on a multistring model, have been given in Sect. 4.6. Figure 8.11 shows the relative probabilities for the production of electrons from Si (at room temperature) calculated by this method, assuming a bulk-like surface structure and the fully stripped charge state, i.e. using (4.29) for the screening length. The latter assumption is consistent with the observed Z_1^2 scaling, discussed in Sect. 8.1. In the calculation of the encounter probabilities, the two-dimensional density of Si L-shell electrons (viewed along Si⟨100⟩) obtained from atomic wave functions, given by Froese Fischer [114], was used. Also, an impact-parameter dependence of Si L-shell ionization based on semi-classical calculations, given by Hansteen et al. [59], was used in the calculation of ionization probabilities.

We see in Fig. 8.11 that the ionization probability is higher than the encounter probability, indicating that the shadowing effect is weaker for the former. This is because ionization can occur even in a distant collision with shell electrons, in contrast to an encounter event. It is therefore likely that the encounter and ionization probabilities will provide an underestimate and an overestimate, respectively, of the effective surface layer thickness. The probabilities decrease with depth to minimum values at 60–70 Å, and increase gradually to ~0.1 at 150–175 Å as the ions approach the neighboring ⟨100⟩ rows of atoms. The effective surface layer thickness is obtained by integrating the probabilities from the surface to a depth equal to $D = 170$ Å. The effective surface layer thicknesses thus obtained for the encounter and ionization probabilities, $t_{\rm enc}$ and $t_{\rm ion}$, respectively, are summarized in Table 8.1. Approximately 80% of the value of each of $t_{\rm enc}$ and $t_{\rm ion}$ results from integration up to a depth of ~70 Å, which corresponds to the first glancing collisions between the ions and ⟨100⟩ rows, while ~20% of the value results from the second set of glancing collisions with the neighboring ⟨100⟩ rows. Therefore, the probabilities due to the second set of glancing collisions make only a small contribution to the observed yield. The experimental effective layer thickness of 35 ± 5 Å lies approximately in the range 27–34 Å, i.e. between the underestimated and overestimated values for the unshadowed Si L-shell electrons, taking account of a possible ambiguity of ~1.4 Å, the (100) interplanar spacing, in the estimated values. It must be pointed out that the estimated values would increase at most by ~5 Å, i.e. the ⟨100⟩ interatomic spacing, if an enhanced value of the thermal vibration amplitude of Si surface atoms was taken into account in the simulations.

Table 8.1. Experimental values of the effective escape length D and the effective surface layer thickness $t_{\rm exp}$ for Si⟨100⟩, together with the calculated values $t_{\rm enc}$ and $t_{\rm ion}$ [84, 85]

Projectile	Electron energy (keV)	D (Å)	$t_{\rm exp}$ (Å)	$t_{\rm enc}$ (Å)	$t_{\rm ion}$ (Å)
92.5 keV/u He^{+}	0.72–0.80	170 ± 5	35 ± 5	27	34
3.5 MeV/u O^{8+}	3.8–4.8	850 ± 40	175 ± 15	130	210
3.5 MeV/u O^{8+}	7.6–9.6	1810 ± 120	264 ± 18	215	388
3.5 MeV/u O^{8+}	8.6–10.8	2390 ± 170	242 ± 17	239	431
8 MeV/u H^{+}	7.6–9.8	2080 ± 30	194 ± 15	~130	
8 MeV/u He^{2+}	7.6–9.8	2080 ± 30	266 ± 15	~180	

We now discuss the behavior of the 0.36–0.40 keV yield in Fig. 8.9. Since the value of W at $x = 0$ is 0.53, the value of D in this case is estimated as

$D = 35/0.53 = 66$ Å by applying (8.2), assuming that the effective surface layer thickness for the channeling case is the same as for the 0.72–0.80 keV yield. The values of the overlayer thicknesses of 30, 60, and 90 Å plus the effective surface layer thickness of 35 Å are 65, 95, and 125 Å, respectively. These values are comparable to or larger than $D = 66$ Å and, accordingly, the effective surface layer thickness in this case cannot be given by the sum of the thickness of the overlayer and the intrinsic layer thickness of 35 Å, unlike the case noted earlier. Instead, the effective surface layer thickness is given by the integrated probability from the overlayer surface to a depth of 66 Å. The values thus obtained for the 30 Å overlayer are 50 and 54 Å for the encounter and ionization probabilities, respectively, from which $W = 50/66 = 0.76$ and $54/66 = 0.82$ are obtained. These values account roughly for the observed value of $W = 0.72$ (Fig. 8.9). For the 60 and 90 Å overlayers, however, similar estimates lead to effective surface layer thicknesses of 66 Å, i.e. $W = 1$. In fact, the value of D used in the present estimate must be interpreted as a mean value, especially for the case of $W \simeq 1$. The observed values of $W < 1$ for the 60 and 90 Å overlayers, which reflect the shadowing effect in a deep region, can be attributed to the escaped electrons that have passed through a surface layer thicker than 66 Å.

It is necessary to discuss dechanneling of the ions after they have passed through the noncrystalline overlayer. Dechanneling in an overlaid crystal was comprehensively studied by Lugujjo and Mayer [115], who presented a plot of the enhancement of the minimum ion-backscattering yield $\Delta\chi_m$ for a crystal as a function of the reduced critical angle and the reduced overlayer thickness.[3] For a thin Si overlayer as in the present case, we obtain $\Delta\chi_m/x \simeq 5.0 \times 10^{-4}\,\text{Å}^{-1}$ for Si⟨100⟩ from this plot. For a crystal without an overlayer, the channeled fraction not contributing to the electron yield is given by $1 - W_0$, where W_0 is the normalized ⟨100⟩ yield for the crystal. Accordingly, the dechanneling-assisted electron yield ΔW is approximately given by

$$\Delta W = (1 - W_0)\Delta\chi_m \,. \tag{8.3}$$

Since $W_0 = 0.205$ for the 0.72–0.80 keV yield, (8.3) leads to $\Delta W = 0.012$, 0.024, and 0.036 for the 30, 60, and 90 Å overlayers, respectively. Rigorously, ΔW should be subtracted from the measured value of W prior to producing a plot such as that in Fig. 8.9 to determine the effective surface layer thickness and the effective escape length. However, this correction makes only a negligible change in the values of $t = 35 \pm 5$ Å and $D = 170 \pm 5$ Å that are determined. For the 0.36–0.40 keV yield, we obtain in a similar manner $\Delta W = 0.007$, 0.014, and 0.021 for the 30, 60, and 90 Å layers, respectively. Again, each of these values is negligibly small compared with the observed value of W.

[3] A plot for a wider parameter range is given in [43].

8.4.2 Thick Unshadowed Layers

Figure 8.4, shown in Sect. 8.2, compares energy spectra of secondary electrons induced by 3.5 MeV/u O^{8+} for Si with and without a noncrystalline Si overlayer of 235 Å thickness under $\langle 100 \rangle$ and random incidence conditions on Si. Experimentally, the presence of the overlayer causes no change in the random spectrum. In contrast, it is clearly seen that, over the whole energy range, the Si overlayer enhances the $\langle 100 \rangle$ yield. Figure 8.12 shows the x dependence of the measured values of W for $\langle 100 \rangle$ obtained using energy windows of 3.8–4.8, 7.6–9.6, and 8.6–10.8 keV for overlaid crystals; this indicates a linear dependence of W on x. It should be noted that W is effectively constant within each of the three energy windows.

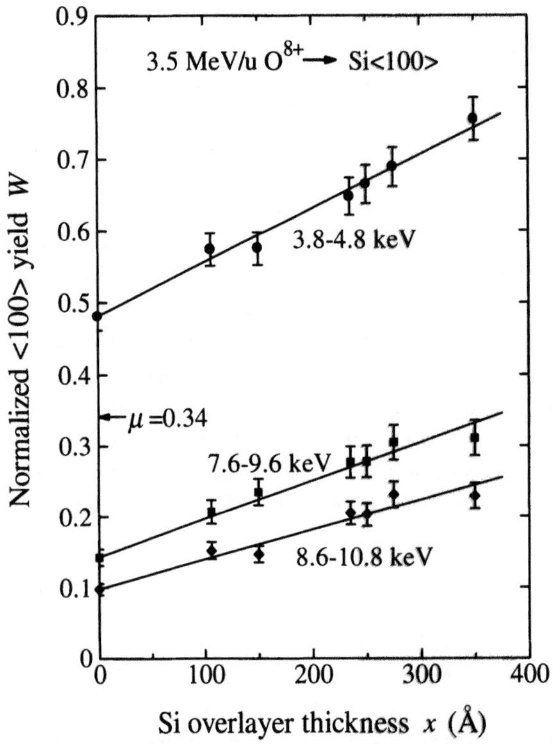

Fig. 8.12. Dependence of W on the thickness of the noncrystalline Si overlayer x, measured for 3.5 MeV/u O^{8+}. The experimentally determined value of $\mu = 0.34$ represents the partial electron yield produced mainly from the Si valence electrons [85]

Analyses similar to those in Sect. 8.4.1 can be applied to the electron yields at 7.6–9.6 and 8.6–10.8 keV since these energies are higher than $E_B = 7.6$ keV. However, a modified analysis is required for the 3.8–4.8 keV yield, and will be presented in Sect. 8.4.3. For the former case, the effective surface layer thicknesses, i.e. the values of x extrapolated to $W = 0$ in Fig. 8.12, and the effective escape lengths determined by extrapolation of the solid lines to $W = 1$ are summarized in Table 8.1. These values are larger than those for the 0.72–0.80 keV yield induced by 92.5 keV/u He^+ by a factor of ~7 for the effective target thicknesses and by a factor of 10–14 for the effective escape lengths, indicating that the electron yield for 3.5 MeV/u O^{8+} is less sensitive to the crystal structure near the surface. Also, (8.2) holds for the intrinsic values of W at $x = 0$.

The maximum energy of the electrons which contribute to the observed yield can be estimated in a similar manner to the case in Sect. 8.4.1, using the observed values of D and $S \approx 0.5\,\text{eV/Å}$ for electrons in the range 8–11 keV [97]. It is therefore concluded that electrons produced in the energy ranges 7.6–10.5 and 8.6-12.0 keV contribute to the observed yields at 7.6–9.6 and 8.6–10.8 keV, respectively.

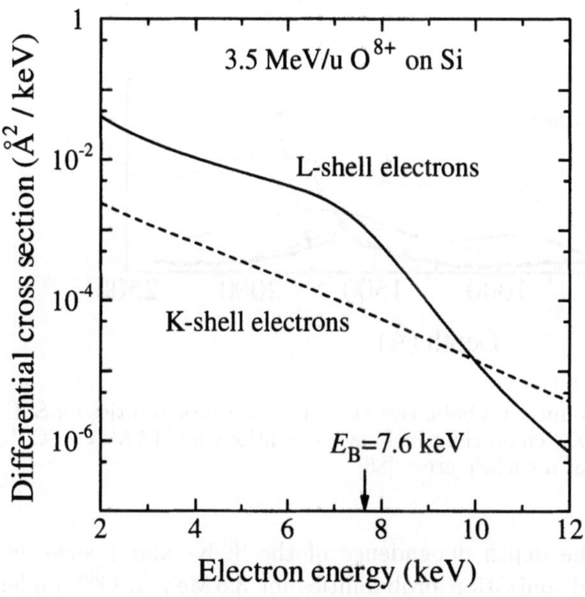

Fig. 8.13. Differential cross sections $\mathcal{Y}_2(E_e)$, given by (3.22), for recoil of K- and L-shell electrons from a Si atom by impact of 3.5 MeV/u O^{8+}. Note that the number of electrons in each shell has been taken into account [85]

Figure 8.13 shows the theoretical differential cross sections $\mathcal{Y}_2(E_e)$, given by (3.22), for recoil of Si K- and L-shell electrons produced by 3.5 MeV/u O^{8+}. It should be noted that in a Si crystal the ions remain fully stripped in the electron production processes in both the channeling and the random case, as will be demonstrated in Chap. 9. The ratio Λ of the integrated value of $\mathcal{Y}_2(E_e)$ over the energy range of interest for K-shell electrons to that for L-shell electrons can be used as a rough measure of the relative K-shell contribution to the observed electron yield. In fact, $\Lambda = 0.14$ in the energy range 7.6–10.5 keV, while $\Lambda = 0.41$ in the range 8.6–12.0 keV, implying that K-shell electrons contribute more effectively to the higher-energy yield. A more refined treatment must include the angular distribution of the recoiled electrons and the dependence of the electron–Si scattering cross section on the electron energy.

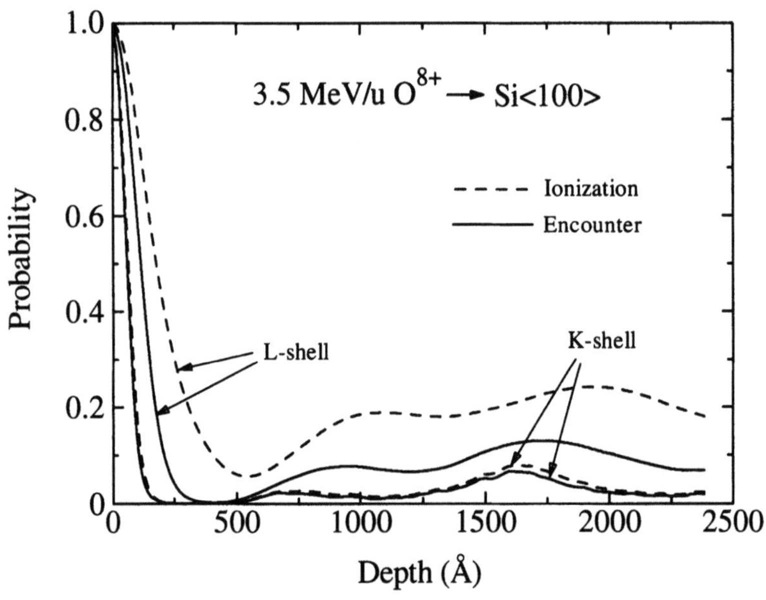

Fig. 8.14. Calculated encounter probabilities and ionization probabilities for Si L-shell electrons under Si⟨100⟩ channeling incidence conditions for 3.5 MeV/u O^{8+}, normalized to the values for a surface atom [85]

Figure 8.14 shows the depth dependence of the Si K- and L-shell encounter probabilities and ionization probabilities for 3.5 MeV/u O^{8+} under ⟨100⟩ channeling incidence conditions, calculated in a similar manner to the case shown earlier in Fig. 8.11. The unshadowed surface layer for the Si L shell seen in the range 0 to ~500 Å is thicker than for 92.5 keV/u He^+ (Fig. 8.11) by a factor of ~6 for the L-shell encounter and ~9 for the L-shell ionization. The former factor can be explained in terms of the trajectory scaling parameter. The ratio of the parameter $V_1\sqrt{M_1/Z_1}$ for 3.5 MeV/u O^{8+} to that for 92.5 keV/u He^+ is equal to 6.15. Trajectory scaling is not applicable to the ionization probabilities because the impact parameter dependence of the ionization based on the semiclassical model adopted in the calculations depends inherently on the ion velocity. The calculated curves for the Si K shell predict a thinner unshadowed layer than for the L shell by a factor of 0.3–0.5. The effective surface layer thicknesses for the Si K and L shells, t_K and t_L, respectively, are obtained by integrating the calculated probabilities from the surface to a depth equal to the value of D. The effective surface layer thickness in this case is approximately given by

$$t = (t_L + \Lambda t_K)/(1 + \Lambda) . \tag{8.4}$$

In Table 8.1, the values of t obtained for the encounter and ionization probabilities, t_{enc} and t_{ion}, respectively, are summarized. The observed effective surface layer thickness of 264 ± 18 Å at 7.6–9.6 keV lies between the calculated values for encounter (215 Å) and ionization (388 Å), corresponding to underestimated and overestimated values, respectively. Similarly, the observed value of 242 ± 17 Å at 8.6–10.8 keV is certainly in the range 239–431 Å. It is noteworthy that the underestimated value of 239 Å for the 8.6–10.8 keV yield is replaced by the value $t_L = 282$ Å if the K-shell contribution is ignored, but this fails to account for the observed value. Furthermore, the use of the L-shell ionization probability considerably overestimates the effective surface layer thicknesses for the electron yield, by a factor of ~1.5 for the yield at 7.6–9.6 keV and ~1.8 at 8.6–10.8 keV, in contrast to the case of ~100 keV/u ions.

8.4.3 Shadowing for All Electrons in a Crystal

The electron yield at 3.8–4.8 keV originates not only from inner-shell but also from valence electrons in Si since this energy range lies below $E_B = 7.6$ keV. In this case the yield component μ, introduced in Sect. 8.3, must be estimated. A value of $\mu = 0.344 \pm 0.03$ is obtained from (8.1) using the observed values of $W = 0.48$ (Fig. 8.12) and $W_p = 0.44 \pm 0.01$ (see Fig. 9.5). The interpretation of the μ value obtained will be given in Sect. 8.5.

The W versus x line experimentally obtained for the 3.8–4.8 keV yield, shown in Fig. 8.12, gives $-x = 175 \pm 15$ Å when it is extrapolated to $W = \mu = 0.344$. This value should be equal to the effective surface layer thickness for the first glancing collisions with the $\langle 100 \rangle$ rows, which can be approximately obtained by integrating the calculated encounter or ionization probabilities for the Si K and L shells from the surface to a depth of ~500 Å, where the first glancing collisions are completed, as seen in Fig. 8.14. In the electron energy range of interest (more precisely 3.8–5.3 keV if the energy loss suffered by the escaped electron is taken into account), the calculated differential cross section for the Si K shell is less than for the L shell by a factor as small as ~0.06 (Fig. 8.13). It is therefore reasonable that the observed value of 175 ± 15 Å lies between the values calculated for the L-shell encounter and ionization probabilities, i.e. 130 and 210 Å, corresponding to an underestimate and an overestimate, respectively. The effective escape length in this case is estimated as $D = 850 \pm 40$ Å by extrapolation of the W versus x line to

$W = 1$. Obviously, in the present case, the expression corresponding to (8.2) for $x = 0$ is

$$W = \mu + (1 - \mu)t/D\,. \tag{8.5}$$

Similar experiments and analyses for 8 MeV/u H^+ and He^{2+} have been reported [84]. In this case, the electron yield was measured in the range 7.6–9.8 keV, which is roughly a factor of 0.5 lower than $E_B = 17.4$ keV. All the analysis results for ions in the MeV/u energy range are summarized in Table 8.1.

The effect of the dechanneling of ions after they have passed through the noncrystalline overlayer can be estimated in a similar manner to the case discussed in Sect. 8.4.1, on the basis of the calculations by Lugujjo and Mayer. In the high-velocity case, the dechanneling effect again gives rise to only a minor influence on the electron emission under channeling incidence conditions.

8.4.4 Hetero-Overlayers

High-energy shadowing experiments using crystals covered with noncrystalline hetero-overlayers provide further information about the production and escape processes in ion-induced electron emission. In the following, measurements and analyses performed with Si and Ge crystals covered with deposited hetero-overlayers are reviewed [68].

The experimental conditions were similar to the case discussed in Sect. 8.4.2. Figure 8.15 shows energy spectra of secondary electrons induced by 3.5 MeV/u O^{8+} for Si with and without an Ag overlayer 48 Å thick, under Si⟨111⟩ channeling and random incidence conditions. The relative energy resolution (acceptance) of the spectrometer was $\Delta E/E \simeq 7\%$. The spectrum for a noncrystalline Ag specimen is also shown for comparison. Figure 8.16 shows similar spectra for an Au overlayer 98 Å thick on Ge, measured under Ge⟨111⟩ channeling and random incidence conditions.

In Fig. 8.15, the Ag overlayer enhances the ⟨111⟩ yield over the whole energy range, by a factor of ~3 at 8 keV, for example. In contrast, the overlayer gives rise to only small differences in the random spectra (typically within ~10% at 8 keV). Generally, differences in the transport processes corresponding to the different atom species should cause differences in the electron yield. The insensitivity of the random electron yield to the hetero-overlayer indicates that the overlayer is thin enough not to affect the overall transport processes of the electrons. In fact, the energy losses of ~8 keV electrons are 0.5 and 1.5 eV/Å in Si and Ag, respectively [97], which causes an excess energy loss of only ~0.05 keV in the Ag layer. The results for an Au overlayer on Ge, shown in Fig. 8.16, are essentially similar to the case of the Ag layer on Si; the excess energy loss of 8 keV electrons in the Au overlayer is only ~0.1 keV. In these cases, therefore, the overlayers play approximately the role

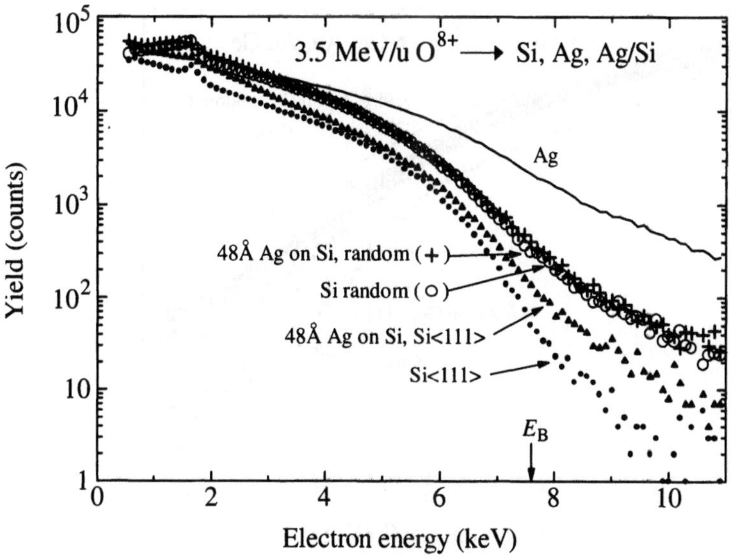

Fig. 8.15. Energy spectra of secondary electrons induced by 3.5 MeV/u O^{8+}, measured for Si, Ag-covered Si, and polycrystalline Ag at a backward angle of 180° [68]

of an electron source on the crystal surface that does not affect the electron transport processes in the target.

In this work, the authors focused on the electron yield in an energy window of 7.6–9.6 keV, which is one of the energy ranges of interest discussed in Sect. 8.4.2. In this energy range, the yield stems mainly from L-shell electrons in the Si target and from M-shell electrons in the Ge target, according to an estimate by the binary-encounter calculations demonstrated in Sect. 8.4.2. For an analysis of the enhanced electron yield due to the overlayers, the ratio of the yield for the overlaid crystal to the random yield for the crystal without an overlayer was measured. The ratios, i.e. the normalized yields, W_c and W_r obtained in this manner for the channeling and random cases, respectively, are discussed in the following analysis.

We first consider the random case. For overlaid atoms of higher atomic number than the crystal atoms, the thin overlayer typically increases the number of keV electrons that recoil in the forward direction. This is expected to increase the normalized yield by $\Delta W_r = W_r - 1$ through the backscattering of the keV electrons in both the crystal and the overlayer. ΔW_r includes a fraction proportional to the thickness of the overlayer L, which results from electron backscattering in the crystal. The remaining fraction of ΔW_r results from electron backscattering in the overlayer, where the atoms have a larger backscattering cross section than do the crystal atoms. In this case, the electrons that are recoiled (in the forward direction) by the ion at a depth z ($< L$) can be backscattered by the overlaid atoms with a probability

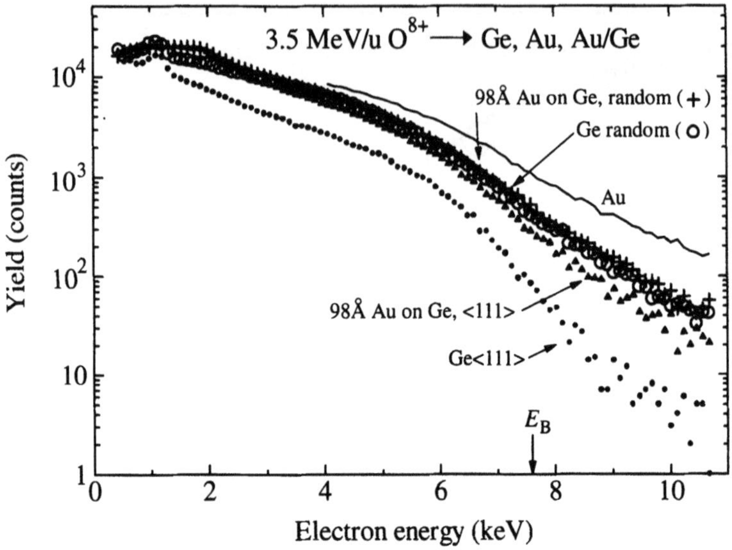

Fig. 8.16. Energy spectra of secondary electrons induced by 3.5 MeV/u O^{8+}, measured for Ge, Au-covered Ge, and polycrystalline Au [68]

proportional to the distance $L - z$. It follows that the increase in the electron yield due to backscattering in the overlayer is proportional to the integral

$$\int_0^L (L - z)\,\mathrm{d}z = L^2/2\,.$$

Therefore, this fraction of ΔW_r should have an L^2 dependence. Taking into account backscattering in both the crystal and the overlayer, the dependence of W_r on L is given by

$$W_\mathrm{r} = 1 + \Delta W_\mathrm{r} = 1 + p_1 L + \ell L^2\,, \qquad (8.6)$$

where the parameters p_1 and ℓ depend on the differences in the ion–electron and electron–atom scattering cross sections in the overlayer and crystal. The present analysis can be extended to the case in which the overlaid atoms are lighter than the crystal atoms by allowing negative values of p_1 and ℓ. Rigorously, the L dependence in (8.6) is correct only for an infinitely thin overlayer. However, its applicability to finite values of L is expected from the fact that the thin overlayer in the present case can be regarded merely as an electron source on the crystal surface, as discussed earlier.

Figures 8.17 and 8.18 show measured values of W_r plotted against L for Ag and Au overlayers on Si and Ge. Measurements on bulk Ag and Au give upper limits on the values of W_r at $L = \infty$, which are approximately 7.1, 15, 1.4, and 3.0 for Ag on Si, Au on Si, Ag on Ge, and Au on Ge, respectively. For Ag overlayers on Ge, the small value of the upper limit on W_r accounts

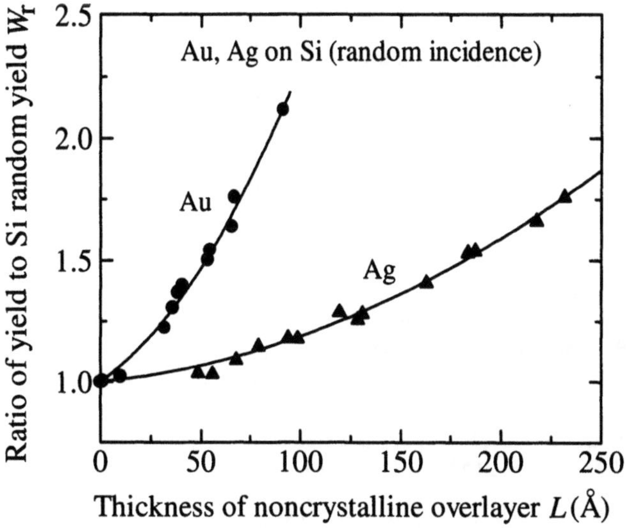

Fig. 8.17. Dependence of the random yield W_r, normalized to the Si random yield, on the thickness L of noncrystalline Au and Ag overlayers on Si. The *solid curves* shows fitted curves obtained from (8.6): $p_1 = 6.4 \times 10^{-3}\,\text{Å}^{-1}$, $\ell = 6.5 \times 10^{-5}\,\text{Å}^{-2}$ for the Au overlayer, and $p_1 = 8.2 \times 10^{-4}\,\text{Å}^{-1}$, $\ell = 1.1 \times 10^{-5}\,\text{Å}^{-2}$ for the Ag overlayer [68]

for the weak dependence of W_r on L. The solid curves in Figs. 8.17 and 8.18 are fitted curves obtained using (8.6), with p_1 and ℓ being used as fitting parameters. The values of the two parameters obtained will be discussed later, in comparison with the channeling case.

Next, we discuss the channeling case. The effective surface layer thickness is increased by the noncrystalline overlayer. Also, there must be enhanced backscattering from the overlayer, as was discussed earlier for the random case. Accordingly, the dependence of W_c on L can be expressed as

$$W_c = W_0 + (r_1 + s_1)L + \ell L^2 , \tag{8.7}$$

where W_0 is the channeling yield for the crystal without an overlayer, and the parameters r_1 and s_1, which take positive values, represent the effective increase in the number of unshadowed atoms and the dechanneling effect after the ions have passed through the thin overlayer, respectively. It should be pointed out that $r_1 > p_1$, since r_1 results from the net increase in the number of unshadowed atoms, while p_1 stems from the replacement of the atom species in the region $0 \le z \le L$. As a special case, $p_1 = \ell = 0$ for an overlayer of the same atom species and atom density as in the crystal.

The experimental results for the Si$\langle 100\rangle$ and Ge$\langle 111\rangle$ channeling cases are shown in Figs. 8.19 and 8.20, respectively. The solid curves are fitted curves obtained using (8.7) with $r_1 + s_1$ and ℓ as fitting parameters. We see that

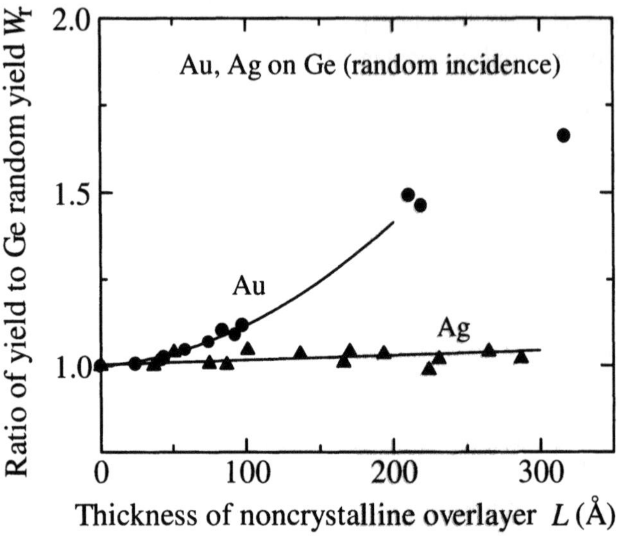

Fig. 8.18. Dependence of W_r on L of noncrystalline Au and Ag overlayers on Ge. The *solid curves* show fitted curves obtained from (8.6): $p_1 = 3.0 \times 10^{-4}\,\text{Å}^{-1}$, $\ell = 8.9 \times 10^{-6}\,\text{Å}^{-2}$ for the Au overlayer, and $p_1 = 1.1 \times 10^{-4}\,\text{Å}^{-1}$, $\ell \simeq 0\,\text{Å}^{-2}$ for the Ag overlayer [68]

in every case W_c increases linearly with increasing thickness of the overlayer, although a slight contribution of the second derivative $d^2W_c/dL^2 = 2\ell\ (>0)$ can be seen. This is contrast to the random case, shown in Figs. 8.17 and 8.18, in which no slope p_1 of the curve can be clearly recognized at $L = 0$, but the L^2-dependent behavior can be clearly seen. In the channeling case, therefore, electron backscattering processes not in the thin overlayer but in a deeper region are mainly responsible for the observed electron yield. This also demonstrates the escape processes of the binary-encounter electrons, as discussed in Sect. 6.1.

While the effective escape length for the electron yield for Si is known to be $D = 1810\,\text{Å}$, as shown in Table 8.1, the effective escape length for Ge has not been estimated yet. According to Fig. 8.18, Ag atoms in an overlayer on Ge provide approximately the same contribution to the electron emission as Ge atoms. It follows that an extrapolation procedure applied to the W_c versus L line for Ag in Fig. 8.20 should give rough values of t for the $\langle 111 \rangle$ incidence and of D for a Ge crystal, taking into account the fact that the atom density of Ag is higher than that of Ge by a factor of ~1.3. The values of t and D obtained in this manner are $t = 156 \pm 15\,\text{Å}$ and $D = 1033 \pm 93\,\text{Å}$, respectively. The value of D for Ge is less than the value of $D = 1810\,\text{Å}$ for Si by a factor of ~0.6. Note that the thickness values of the overlayers used in these experiments are much less than the values of D either for Si or for Ge

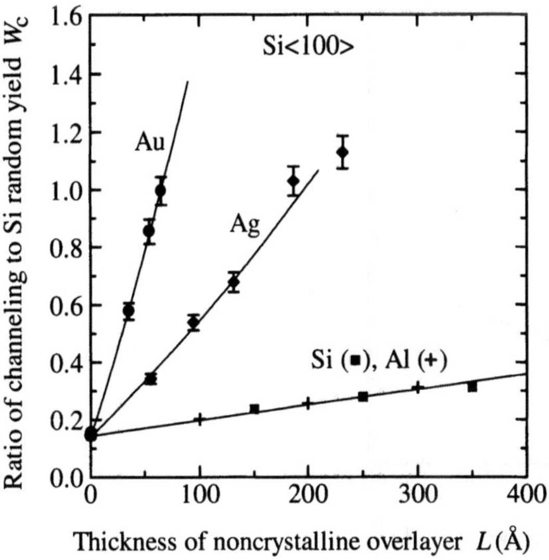

Fig. 8.19. Dependence of the channeling yield W_c for Si⟨100⟩, normalized to the Si random yield, on the thickness L of noncrystalline Au, Ag, Si, and Al overlayers. The *solid curves* show fitted curves obtained from (8.7): $r_1 + s_1 = 1.1 \times 10^{-2}\,\text{Å}^{-1}$, $\ell = 4.2 \times 10^{-5}\,\text{Å}^{-2}$ for the Au overlayer; $r_1 + s_1 = 3.0 \times 10^{-3}\,\text{Å}^{-1}$, $\ell = 1.0 \times 10^{-5}\,\text{Å}^{-2}$ for the Ag overlayer; and $r_1 + s_1 = 5.4 \times 10^{-4}\,\text{Å}^{-1}$, $\ell = 0\,\text{Å}^{-2}$ for the Si and Al overlayers [68]

cystals, which is a necessary condition for the validity of the present analysis of the experimental results.

According to Sect. 8.4.1, the dechanneling-assisted electron yield $s_1 L$ can be written

$$s_1 L = (1 - W_0)\Delta\chi_m \,, \tag{8.8}$$

from which the value of s_1 can be obtained. For example, $W_0 = 0.14$ and $\Delta\chi_m/L = 1.5 \times 10^{-4}\,\text{Å}^{-1}$ for the pair of Si⟨100⟩ and a thin Ag overlayer [43, 115], so that $s_1 = 1.3 \times 10^{-4}\,\text{Å}^{-1}$. Since $r_1 + s_1 = (3.3 \pm 0.4) \times 10^{-3}\,\text{Å}^{-1}$ is obtained from the fitting procedure, the value of r_1 is estimated as $(3.2 \pm 0.4) \times 10^{-3}\,\text{Å}^{-1}$. Furthermore, the value of $\ell = (1.1 \pm 0.1) \times 10^{-5}\,\text{Å}^{-2}$ accounts well for both the channeling and the random data for Ag-covered Si. For all the measurements, as in the case above, the parameter s_1 is negligible in comparison with r_1. The analysis indicates that the enhanced electron yield due to the overlayer results dominantly from an effective increase in the number of recoiled electrons that are incident on the underlying crystal. Only a minor change in the channeling yield, compared with the dominant increase mentioned above, results from the dechanneling caused by the overlayer and from the difference in the electron backscattering cross section between the

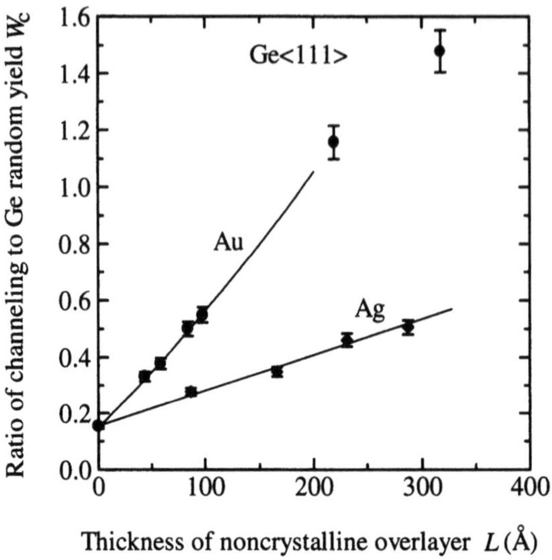

Fig. 8.20. Dependence of W_c for Ge⟨100⟩ on the thickness L of noncrystalline Au and Ag overlayers. The *solid curves* show fitted curves obtained from (8.7): $r_1 + s_1 = 3.2 \times 10^{-3}$ Å^{-1}, $\ell = 1.0 \times 10^{-5}$ Å^{-2} for the Au overlayer, and $r_1 + s_1 = 8.7 \times 10^{-4}$ Å^{-1}, $\ell = 0$ Å^{-2} for the Ag overlayer [68]

overlayer and crystal atoms. However, the latter appreciably influences the observed spectrum under random incidence conditions.

The value of r_1 divided by the atom density in the overlayer represents the relative efficiency per overlaid atom for production of the electron yield. Figure 8.21 shows the dependence of the relative efficiency thus obtained, which is normalized to the value for an overlaid Si atom, on the atomic number of the overlaid atom for a Si target. The ratio of the efficiency for Au to that for Ag, ~3 for a Si target, has been reproduced for overlaid Ge crystals. This again indicates that the overlayer can be regarded as an electron source on the crystal surface, which provides electrons that recoil into the crystal.

The influence of the charge state of the ion will be briefly considered. Screening of the ion's nuclear charge by bound electrons in the inner shells enhances the electron recoil cross section for small recoil angles (hard recoils), which is responsible for production of the keV electron yield, as will be discussed in Chap. 9. In the energy range of the O ions used in the experiments described above, we may assume that the most probable charge state after passing through the overlayer is O^{7+} or O^{8+} [116, 117]. A rough estimate indicates that the electron yield for O^{7+} in Si or Ge is enhanced by a factor

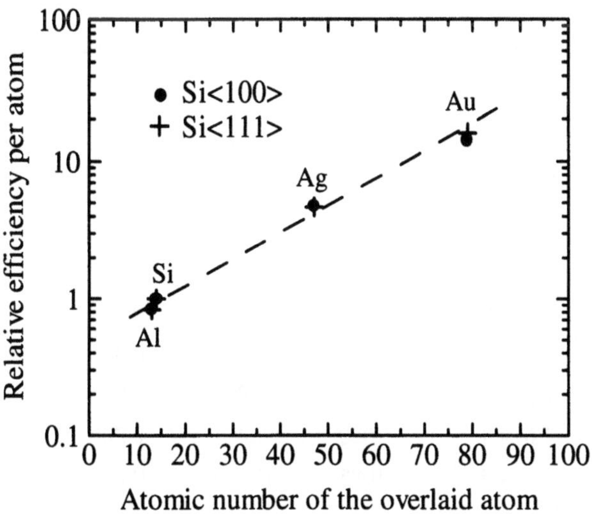

Fig. 8.21. Relative efficiency per atom for production of the electron yield plotted as a function of the atomic number of the overlaid atom, obtained from measurements for Si⟨100⟩ and Si⟨111⟩. The uncertainty in the relative efficiency is roughly ±20% [68]

of at most 1.1,[4] relative to the case of O^{8+}. This implies that the value of W_0 must be replaced by a corrected value lying in the range W_0 to $1.1W_0$. However, such corrections make only a negligible change to the values of the parameters r_1 and s_1 determined.

8.4.5 Negligible Intrinsic Dechanneling

The set of data for the effective escape length, summarized in Table 8.1, provides further information about ion-induced electron emission under channeling incidence conditions. The quantity of interest is the energy loss ΔE_1 of the ion when traveling a distance equal to D. Table 8.2 shows the estimated values of ΔE_1 and the ratio of ΔE_1 to the incident energy of the ion $\Delta E_1/E_1$. Clearly, ΔE_1 is negligibly small compared with E_1.

Dechanneling of fast ions in defect-free crystals can usually be observed in Rutherford backscattering spectra of the ions as an effective increase of the backscattering yield with increasing depth [15, 43]. Experimentally, dechanneling becomes noticeable for ions that suffer an energy loss much more than ~10% of the incident energy. We may conclude from the above estimate of the energy loss that the intrinsic dechanneling in a defect-free crystal is of minor importance in electron emission under channeling incidence conditions.

[4] The factor 1.1 results from a factor of 1.07 (due to the reduced shadowing effect) multiplied by 1.04 (due to the enhanced recoil effect) [68]; see Sects. 9.2 and 9.3.

Table 8.2. Estimated energy losses ΔE_1 of an ion after passing through a distance equal to D, and the ratio $\Delta E_1/E_1$. The values of the stopping power used were taken from [109]

Projectile	Target	Electron energy (keV)	D (Å)	ΔE_1 (keV)	$\Delta E_1/E_1$ (%)
92.5 keV/u He^+	Si	0.72–0.80	170	7.1	1.9
3.5 MeV/u O^{8+}	Si	3.8–4.8	850	91	0.16
3.5 MeV/u O^{8+}	Si	7.6–9.6	1810	200	0.36
3.5 MeV/u O^{8+}	Si	8.6–10.8	2390	250	0.45
8 MeV/u H^+	Si	7.6–9.8	2080	1.9	0.24
8 MeV/u He^{2+}	Si	7.6–9.8	2080	7.4	0.23
3.5 MeV/u O^{8+}	Ge	7.6–9.6	1033[a]	165	0.30

[a] Estimated from the Ag overlayer; see Sect. 8.4.4.

8.5 Unshadowed Crystal Electrons

The value of μ determined from (8.1) is expected to reflect the spatial distribution of valence and loosely bound electrons in the crystal lattice [44]. In a Si crystal, K- and L-shell electrons can be shadowed, so that μ corresponds to the valence electrons. For 3.5 MeV/u ions, the differential cross section $\mathcal{Y}_2$ given by (3.22) for Si K-shell electrons is a factor of ~0.2 smaller than that for Si L-shell or valence electrons in the electron energy range of interest, i.e. 4–6 keV (Fig. 8.13). Furthermore, the L-shell and valence electrons have approximately equal values of this differential cross section. We may therefore assume that, neglecting the two K-shell electrons, $14 - 2 = 12$ electrons per Si atom contribute to the observed random yield. From the value of $\mu = 0.344$ estimated in Sect. 8.4.3, we obtain $12\mu \simeq 4.1$, which is approximately equal to the number of Si valence electrons. It should be noted that the above value of μ, obtained using 3.5 MeV/u ions, has been reproduced well ($\mu = 0.348 \pm 0.02$) using 8 MeV/u H^+ and He^{2+} [84].[5]

A similar analysis of the unshadowed electrons in Ge using 2.5 MeV/u ions has been reported [69]. In this case, μ = 0.246 and 0.198 for Ge⟨100⟩ and Ge⟨110⟩, respectively (Table 9.1). From these values, the effective numbers of unshadowed electrons are estimated as $(32 - 10)\mu \simeq 5.4$ and 4.4 (within 5% uncertainty) for ⟨100⟩ and ⟨110⟩, respectively, neglecting K- and L-shell electrons for a similar reason to that for Si. For the ⟨100⟩ direction, the larger value of 5.5 for Ge than that of 4.1 for Si may possibly be due to the unshadowed 3d electrons, which are widely distributed in the Ge lattice.

[5] The results of the earlier experiments and analysis related to the unshadowed electrons [44] have been refined by recent studies [84,85].

It is notable that the value of μ for $\langle 110 \rangle$ is smaller than for $\langle 100 \rangle$. A similar directional dependence of μ has been observed for Si [118]. Evidently, the value of μ must include the electron yield produced by the second- and higher-order glancing collisions between ion and $\langle 100 \rangle$ rows, which can possibly lead to the recoil, for example, of Si K- and L-shell electrons. This can actually be recognized in Fig. 8.14; see the fluctuating probability values at depths deeper than ~500 Å. For more detailed analysis of the values of μ, the spatial distribution of valence electrons within the Si lattice and the ion flux distribution in the crystal channel must be taken into account. An analytical approach might be possible by assuming statistical equilibrium of the channeled ions (Sect. 4.5.4).

Obviously, the parameter μ results from the preferential interaction of the ion beam with valence electrons under channeling incidence conditions. From the viewpoint of ion beam crystallography, the enhanced sensitivity to valence electrons is of technical interest. This sensitivity is in contrast to structural analysis by X-ray diffraction, in which all electrons in the crystal contribute to the diffraction pattern. Therefore, an improved analysis of the binary-encounter yield should provide a high-sensitivity determination of the spatial density of valence electrons in a crystal, for example from a comparison of the values of μ for different axial directions.

A related study has been reported by Hunt et al. [119]. These authors observed enhanced two-photon annihilation of channeled MeV positrons relative to single-photon annihilation, indicating the preferential interaction of positrons with quasi-free valence electrons. Hence these authors proposed a method for the analysis of valence electrons in a crystal by positron annihilation measurements in conjunction with channeling.

8.6 Classical Aspects of High-Energy Shadowing

The classical behavior of ions in gentle collisions with aligned atoms, predicted by Lindhard from the continuum model (see Sect. 4.5.3), has generally been accepted as a fundamental aspect of ion channeling and related phenomena. Therefore, a description of channeling based on the classical trajectories of ions is valid even for relativistic protons [53, 120, 121], for which the Coulomb scattering parameter satisfies $\kappa \ll 1$.[6] Nevertheless, the criterion for the validity of classical trajectories discussed in Sect. 4.5.3 is *conditional* [13] and does not necessarily hold for collisions with one or two atoms. Accordingly, whether or not the wave nature of the projectiles appears in channeling-related phenomena depends on the experimental conditions. For example, observation of diffraction effects in blocking experiments has been attempted by Lottner et al. [123].

[6] The Coulomb parameter in the relativistic case is formally the same as (4.12) with V replaced by the velocity of light [122].

Binary-encounter electron spectroscopy allows one to demonstrate the classical behavior of ions in the high-energy shadowing effect. In the experiments by Kudo et al. [44], the high-energy shadowing effects of equal-velocity $^{2}H^{+}$ and $^{4}He^{2+}$ in MgO and Si crystals were compared. For these ions, the high-energy shadowing is identical in the classical regime because the values of the parameter R_c given by (4.26) are the same; see Sects. 4.4 and 4.5.1. However, the value of κ, defined by (4.12), for $^{2}H^{+}$ is half the value for $^{4}He^{2+}$. At an energy of 8.35 MeV/u, for example, $\kappa = 1.1$ and 2.2 for the scattering of $^{2}H^{+}$ and $^{4}He^{2+}$, respectively, in an MgO crystal (taking the averaged value of $Z_2 = 10$). In an ion–atom scattering event with $\kappa \approx 1$, quantum-mechanical diffraction is enhanced and, accordingly, the Coulomb shadow becomes degraded, as discussed in Sect. 4.2.

Therefore, if the high-energy shadowing effect is classical, the normalized channeling yield W must be the same for equal-velocity $^{2}H^{+}$ and $^{4}He^{2+}$. The diffraction effect should, however, cause a difference in the normalized yield: the value of W for $^{2}H^{+}$ should be larger than for $^{4}He^{2+}$, since the shadowing effect is reduced more in the former case (i.e. for the smaller value of κ) by the stronger diffraction. A similar comparison of the values of W can also be made for equal-velocity ions of ^{10}B, ^{12}C, ^{16}O, etc., if they are fully stripped in the crystal.

Table 8.3 shows a comparison between the values of W for $^{2}H^{+}$ and $^{4}He^{2+}$ of equal velocity, measured under $\langle 100 \rangle$ axial and (100) planar incidence on MgO, together with the values of κ. For each pair, the values of W are equal within an estimated uncertainty of ~3%. Furthermore, the corresponding channeling angular dips for these ions exhibited no discernible difference. In a similar manner, pairs of W values for other ions of equal velocity, as well as for other target crystals, can be iteratively compared to extend the range of κ in which no discernible difference is observed in the values of W. These results indicate that the shadowing effect is independent of κ in the range $\kappa = 1.1$–41. Hence the classical behavior of ions in the high-energy shadowing effect has been experimentally confirmed.

Table 8.3. Comparison between the measured values of W for 8.35 MeV/u $^{2}H^{+}$ and $^{4}He^{2+}$ for channeling incidence on MgO. W was determined in the electron energy range 5–11 keV. The values of κ for the averaged value of $Z_2 = 10$ are also shown [44]

Channel	Projectile	W	κ
MgO$\langle 100 \rangle$	$^{2}H^{+}$	0.51 ± 0.01	1.1
MgO$\langle 100 \rangle$	$^{4}He^{2+}$	0.52 ± 0.01	2.2
MgO (100)	$^{2}H^{+}$	0.74 ± 0.02	1.1
MgO (100)	$^{4}He^{2+}$	0.74 ± 0.02	2.2

For ⟨100⟩ channeling incidence of 8.35 MeV/u $^{2}H^{+}$ on MgO the average diffracted flux behind the surface atoms, including thermally displaced atoms that are fully exposed to the ion beam near the surface, is given by $\mathcal{D}(1.1) = 0.33$, according to Fig. 4.4. After passing through $\mathcal{N}$ atoms, the diffracted flux should be reduced to approximately $[\mathcal{D}(1.1)]^{\mathcal{N}}$, neglecting a possible higher-order contribution due to backscattering of the flux. The above quantity is appreciable only for $\mathcal{N} \lesssim 4$. Therefore, it is hard for the diffracted flux to remain along an atomic row, where scattering very frequently occurs. Clearly, the diffraction effect is negligible in the present case because the high-energy shadowing effect results from the scattering of ions by more than several tens of atoms, according to the numerical calculations.

9. Electron Emission by Partially Stripped Ions

The high-energy shadowing effect of partially stripped ions should be weaker than of fully stripped ions because of the reduced ion–atom interaction due to the short-range (~0.1 Å) screening of the ion's nuclear charge by the bound electrons. This leads to the idea of applying binary-encounter electron spectroscopy to the observation of the charge states of fast ions in a crystal. In this case, however, another screening effect, i.e. the enhanced electron emission in the forward direction, must be taken into account. This chapter is devoted to a review of the studies of charge states of fast ions in crystalline solids [69, 118, 124].

9.1 Charge States of Channeled Ions

When fast ions are incident on a solid, the initial charge state varies towards an equilibrium charge state with increasing penetrating distance, as a result of the balance between electron capture and loss processes in the ion–atom collisions [87, 125]. The equilibrium charge state distributions of fast ions have been determined for various experimental parameters by transmission experiments using thin foils [117, 126–128]. At present, the charge states of ions resulting from ion–solid interactions are hard to predict precisely from theory, and therefore compilation of experimental data is of special importance for a comprehensive understanding of the phenomenon.

Because it affects the balance between capture and loss processes, channeling changes the charge state distribution of ions transmitted through a single crystal. For example, Martin observed a slight increase of the mean charge for channeled ions compared with the random (nonchanneling) case for C and O ions at 5.2–35.4 MeV in Si and Ni [129]. On the other hand, Lutz et al. observed a decrease of the mean charge for 60 MeV I ions in Au [130]. It is notable that for channeled ions the distance required to reach charge equilibrium is longer than in the random case because of the reduced ion–atom interaction. In fact, Datz et al. have observed nonequilibrium charge states (depending on the input charge state) in transmission channeling experiments, which are dependent on the oscillation amplitude of the channeled ions [116]. In this manner, channeling experiments provide unique information about electron loss and capture processes under conditions where the impact parameters of the ion–atom collisions are restricted [15].

It is of physical as well as technical interest to observe the charge states of ions inside a solid, rather than after they have passed through a foil, using the high-energy shadowing technique. Furthermore, the charge states determined by this method will be those of the ions passing close to the aligned atoms so that they effectively produce binary-encounter electrons. It is therefore likely that the charge states in this case will be different from those observed in transmission channeling experiments, in which the charge states of the ions that have passed through the crystal channel are measured.

9.2 Reduced Shadowing Effect

As was discussed in Sect. 4.4, the high-energy shadowing effect is characterized by a single parameter, i.e. the shadow cone radius given by (4.26):

$$R_c = (8Z_1Z_2e^2d/M_1V_1^2)^{1/2} .$$

We may expect from this expression that, for a given channeling direction of a crystal, the normalized channeling yields should be the same for equal-velocity ions if they are fully stripped and have the same Z_1/M_1 value, such as ^{2}H, ^{4}He, ^{10}B, ^{12}C, ^{16}O,..., and, neglecting a small difference, also ^{35}Cl.

Since at channeling incidence most ions undergo only soft collisions with impact parameters larger than the typical inner-shell radius (of the order of 0.1 Å), the ion's nuclear charge is effectively screened when the inner shells of the ion are occupied by bound electrons. Screening reduces the ion–atom interaction, resulting in a reduced shadowing effect compared with the fully stripped case. When the ion's nuclear charge is effectively screened by the inner-shell electrons, Z_1 in (4.26) must be replaced by an effective nuclear charge Z_{eff} which is less than Z_1. For an ion with two bound electrons in the K shell, for example, we may assume $Z_{\text{eff}} \simeq Z_1 - 2$. The reduced shadow cone radius in such a case can be written

$$R_c = (8Z_{\text{eff}}Z_2e^2d/M_1V_1^2)^{1/2} , \tag{9.1}$$

instead of (4.26).

The shadow cone radius cannot be directly related to the high-energy shadow that develops around the aligned atoms, as discussed in Sect. 4.4. Nevertheless, it is reasonable and even intuitive to account for the reduced high-energy shadow in terms of a reduced shadow cone radius due to the screened nuclear charge.

According to Sect. 4.5.1, R_c^{-1} is proportional to the trajectory scaling parameter $V_1\sqrt{M_1/Z_1}$, which is also the scaling factor for the depth evolution of the high-energy shadowing effect. Therefore, the effective thickness of the unshadowed surface layer for partially stripped ions must be a factor of $(Z_1/Z_{\text{eff}})^{1/2}$ thicker than for fully stripped ions of equal velocity.

It should be noted that Z_{eff} for the high-energy shadowing effect does not directly correspond to the effective nuclear charges defined for other atomic processes. For example, an effective nuclear charge for the stopping powers of heavy ions has been introduced for scaling from the values of the stopping power of protons [109]. Accordingly, it includes the screening effect in ion–atom collisions at essentially all impact parameters, in contrast to the case of high-energy shadowing. Furthermore, Z_{eff} will not be necessarily equal to the mean charge or most probable charge determined from foil transmission experiments, since these quantities include a number of loosely bound electrons that seldom screen the ion's nuclear charge in the high-energy shadowing effect.

9.3 Enhanced Recoil Effect

9.3.1 Electron Scattering by a Screened Coulomb Field

The screened interaction potential of partially stripped ions causes a less simple modification to the production of binary-encounter electrons by fully stripped ions. Remarkably, the screening effect modifies the Rutherford recoil cross section, especially at 0°, as shown later. The effect of screening on the binary-encounter peak yield at 0° has been one of the main subjects of zero-degree electron spectroscopy. The dependence of the binary-encounter peak yield on the charge of the ions has been studied experimentally as well as theoretically [7, 131–136]. According to these studies, the binary-encounter peak yield at ~0° increases with decreasing charge of the ions in the present case, i.e. in the MeV/u energy range. Screening of the ion's nuclear charge thus gives rise to enhanced electron recoil, typically in a forward direction. This enhancement generally depends on the ion velocity and recoil angle in a complicated manner, and can be accounted for on a quantum-mechanical basis.

For more detailed explanation of the enhanced recoil effect, elastic scattering of electrons by neutral ions (atoms), i.e. the most screened case, can be considered. On the basis of atomic wave functions, Riley, MacCallum, and Biggs (RMB) calculated electron–atom elastic-scattering cross sections by the partial-wave method and gave these in tabular form [102]. The values of the cross sections σ_{RMB} are given as a function of the scattering angle φ in the rest frame of the atom. The conversion from φ to the angle φ_1 in the laboratory frame, where an electron initially at rest is scattered by the projectile atom, is given by

$$\cos 2\varphi_1 = -\cos\varphi, \quad \text{i.e.} \quad 2\varphi_1 + \varphi = 180° \,. \tag{9.2}$$

In the rest frame of the atom, the incident energy of the electrons is equal to the loss-peak energy E_{L}. For example, $E_{\text{L}} = 1.36$ and $1.91\,\text{keV}$ for the 2.5 and 3.5 MeV/u ions respectively, used in the experiments.

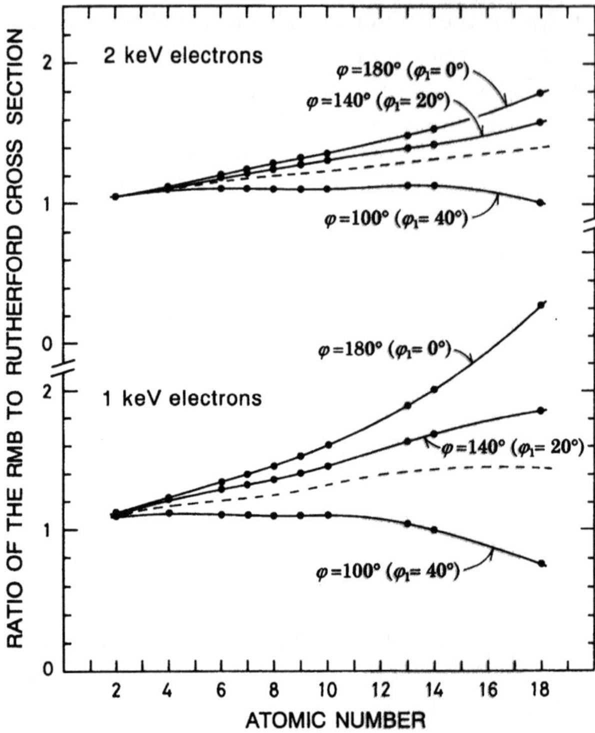

Fig. 9.1. Ratio of the Riley–MacCallum–Biggs cross section to the Rutherford differential scattering cross section for elastic scattering of electrons, plotted as a function of the atomic number of the target atom. The *solid curves* are drawn to guide the eye. The scattering angle φ in the rest frame of the atom can be converted to the emission angle φ_1 in the laboratory frame by (9.2). The *dashed curves* represent the ratios for the integrated cross sections from $\varphi_1 = 0$ to 38° (see text) [118]

Figure 9.1 shows the ratio of σ_{RMB} to the Rutherford differential scattering cross section for 1 and 2 keV electrons as a function of the atomic number of the target atom. For any scattering angle, the ratio depends on the atomic number, i.e. Z_1 in this case, which demonstrates that the differential cross section for a neutral atom is not proportional to Z_1^2. Moreover, the ratio depends on the electron energy and scattering angle. The ratio is greater than unity at $\varphi \simeq 180°$ ($\varphi_1 \simeq 0°$), so that the differential cross section for the screened case is enhanced relative to the Rutherford, i.e. fully stripped, case. The ratio can also become less than unity, as seen for 1 keV electrons scattered by relatively heavy atoms at an angle of $\varphi \simeq 100°$. Such complicated behavior of the electron scattering intensity relative to the unscreened case occurs more pronouncedly at lower incident energies of the electron.

9.3.2 Enhancement Factor

We first consider the choice of the energy of the electrons to be measured. For precise measurements of the binary-encounter yield from light target atoms such as Si, this energy should be chosen between E_L and E_B to obtain intense electron yields. For a crystal consisting of heavy atoms, intense yields can be obtained even at an electron energy higher than E_B. In a typical case, the electron energy for such measurements is chosen at $\sim(E_L+E_B)/2 = (5/8)E_B$, which is equal to the kinetic energy transferred to an electron initially at rest and recoiled at $\varphi_1 = 38°$, according to (3.1). The electrons recoiled at $\varphi_1 < 38°$ have a higher kinetic energy, but can contribute to the measured electron yield after losing energy in their outgoing path. Therefore, the electron yield should have approximately the same charge-state dependence as integrated cross section for $0 \leq \varphi_1 \leq 38°$ $(180° \geq \varphi \geq 104°)$, i.e.

$$\Omega = \int_{104°}^{180°} \sigma_{\mathrm{RMB}}\, 2\pi \sin\varphi \, \mathrm{d}\varphi \, . \tag{9.3}$$

It should be noted that (9.3) is meaningful only when the orbital velocity of the electron is much lower than the ion velocity. In other words, the binding energy of the electron must be small relative to the value of E_L, which is equal to 1.36 and 1.91 keV for 2.5 and 3.5 MeV/u ions, respectively. In the present case, this condition is fulfilled at energies below $\sim E_B$, where the outer-shell (~100 eV binding energies for Si and Ge) and valence electrons mainly contribute to the electron yield.

The ratio of Ω to the integrated Rutherford cross section for the same angular range Ω_R, i.e.

$$\xi = \Omega/\Omega_R \, , \tag{9.4}$$

is shown by the dashed curves in Fig. 9.1. We see that $\xi > 1$ for the whole range of atomic numbers shown in Fig. 9.1, indicating that the electron yield is enhanced by screening of the nuclear charge by bound electrons, relative to the fully stripped case. The values of ξ for 1.36 and 1.91 keV to be used in the analysis can be obtained by interpolation between the values for 1 and 2 keV. For ions of charge q $(0 \leq q \leq Z_1)$, we define the enhancement factor of the electron yield $F(Z_1, q)$ by

$$F(Z_1, q) = 1 + [\xi(Z_1) - 1]\,(Z_1 - q)/Z_1 \, . \tag{9.5}$$

Clearly, $F=1$ for fully stripped ions $(q=Z_1)$, and $F=\xi$ for the neutral case $(q = 0)$. The linear dependence of $F(Z_1, q)$ on q in (9.5) roughly reproduces the experimental charge state dependence of the binary-encounter peaks for F, Si, and Ni ions in the MeV/u range, studied by Sataka et al. [136], for example. Figure 9.2 compares the experimental results for 4 MeV/u Si ions obtained by Sataka et al. with $F(Z_1, q)$ for $\xi = 1.53$, calculated from the data shown in Fig. 9.1.

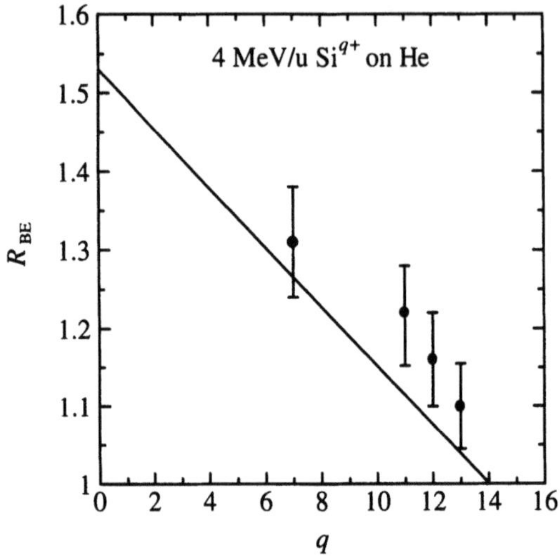

Fig. 9.2. Ratio R_{BE} of the binary-encounter peak intensity for 4 MeV/u $Si^{7,11,12,13+}$ on He to that for Si^{14+}. The linear dependence $R_{BE} = F(14, q)$, adopted in the present model, is shown by the *solid line.* Note that $R_{BE} = \xi = 1.53$ at $q = 0$. Adapted from [136]

9.4 Method of Analysis

As discussed in Sects. 9.2 and 9.3, the channeling electron yield induced by partially stripped ions can be enhanced by two independent effects arising from the screening of ion's nuclear charge, i.e.

- the reduced shadowing effect, which is associated with the reduced value of R_c given by (9.1)
- the enhanced recoil effect, which increases the production of high-energy electron yield, as discussed in Sect. 9.3.2.

Experimentally, the screening effect causes an enhancement of the normalized channeling yield relative to that for fully stripped ions of equal velocity. In this case, the enhanced recoil effect in the random case must be taken carefully into account unless the ions are fully stripped in the crystal, since it affects the normalization of the channeling yield.

For analysis of the difference between the normalized channeling yields for partially and fully stripped ions, the normalized yield for the screened case W_s can be converted to a modified yield W_{mod} by eliminating the enhanced recoil effect, i.e.

$$W_{mod} = \frac{F(Z_1, q_r)}{F(Z_1, Z_{eff})} W_s \,, \tag{9.6}$$

where q_r is the effective nuclear charge for the random case and can be replaced by the most probable charge in the charge equilibrium of ions that have passed through a foil (Sect. 9.5.3).

It follows reasonably that the difference between W_{mod} and the normalized channeling yield for fully stripped light ions W_f ($< W_s$) can be attributed only to the reduced shadowing effect. From the scaling of the ion trajectory with respect to the ion's nuclear charge, as mentioned in Sect. 8.3, we may write

$$Z_{\text{eff}} = \left(\frac{W_f - \mu}{W_{\text{mod}} - \mu} \right)^2 Z_1 , \tag{9.7}$$

where the unshadowed component μ can be obtained from (8.1), using the measured values of W_f and W_p for equal-velocity protons. Z_{eff} can be determined by solving (9.7), which is actually a cubic equation in Z_{eff}.

9.5 Determination of Z_{eff}

9.5.1 Experimental Data

Figure 9.3 shows energy spectra of secondary electrons emitted from a Si crystal target bombarded by 2.5 MeV/u O^{5+} under $\langle 100 \rangle$ channeling and random incidence conditions, and the normalized $\langle 100 \rangle$ channeling yield obtained from the spectra. The observation angle of the electrons was 180° with respect to the beam direction. The relative energy resolution, i.e. the energy acceptance, was chosen to be $\Delta E/E \simeq 10\%$ to obtain a sufficient count rate of the electron signal. The results for Ge$\langle 110 \rangle$ were shown earlier in Fig. 8.5. In these cases, the shadowing effect for all inner-shell electrons is discernible from the marked difference in the normalized yield above and below $E_B = 5.45$ keV. For 2.5 MeV/u ions, the normalized channeling yield W was measured at 3.3 keV, which is between $E_L = 1.36$ keV and $E_B = 5.45$ keV. Similarly, W was measured at 4.4 keV for 3.5 MeV/u ions, for which $E_L = 1.91$ keV and $E_B = 7.63$ keV. The measured electron yield was typically greater than 10^4 counts in both the channeling and the random cases.

The experiments were carried out for most of the charge states of the ions available from three accelerators covering different energy ranges, sometimes using two charge-stripper foils to produce highly stripped ions. The data accumulated in this manner typically include two input charge states of 2.5 MeV/u ions of ^{10}B, ^{12}C, ^{16}O, ^{28}Si, ^{32}S, and ^{35}Cl, and similarly for 3.5 MeV/u ions. Equal-velocity $^1H^+$ and $^4He^{2+}$ ions were also used in the experiments for acquisition of reference data for fully stripped ions in the crystal, since these light ions hardly capture electrons in the target crystal, as seen in Sect. 8.1.

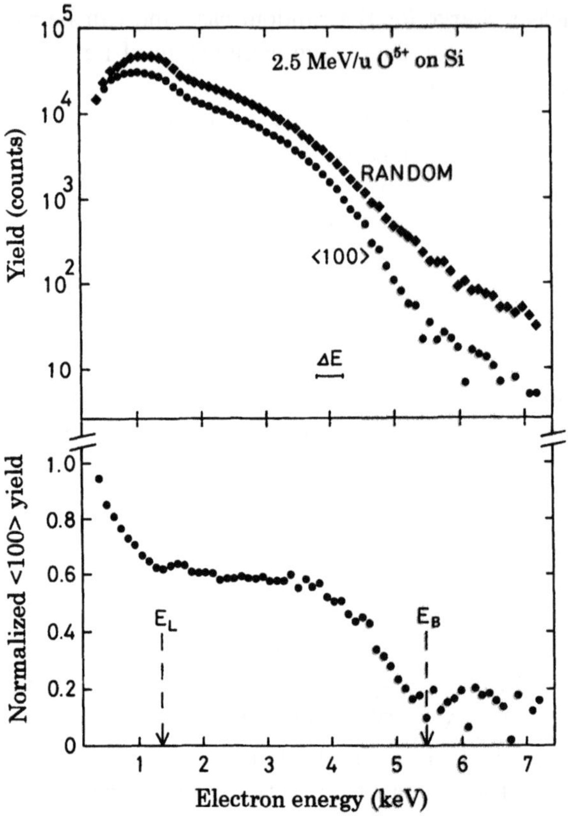

Fig. 9.3. Energy spectra of secondary electrons from Si induced by 2.5 MeV/u O^{5+} under ⟨100⟩ channeling and random incidence conditions, measured at 180°, and the normalized ⟨100⟩ yield [118]

Figure 9.4 shows the values of W as a function of Z_1 which were measured for 3.3 keV electrons induced by the 2.5 MeV/u ions under Si⟨100⟩ and Si⟨110⟩ channeling incidence conditions. Similar results are shown in Fig. 9.5 for 4.4 keV electrons induced by the 3.5 MeV/u ions incident on a Si crystal, and in Fig. 9.6 for 3.3 and 7.7 keV electrons induced by the 2.5 MeV/u ions incident on a Ge crystal. In Figs. 9.4–9.6, the numbers near the plots indicate the input charges of the ions. We see that for each ion species of $Z_1 \geq 6$ the value of W for the lower input charge is greater than for the higher input charge. It is clear from this result that the ions are in nonequilibrium charge states, as is often observed in transmission channeling experiments [116].

9.5.2 Screened and Unscreened Yields

As seen in Figs. 9.4–9.6, the values of W for the fully stripped incident ions are the same for different ions; for example, $W = 0.538$ and 0.435 for Si⟨100⟩

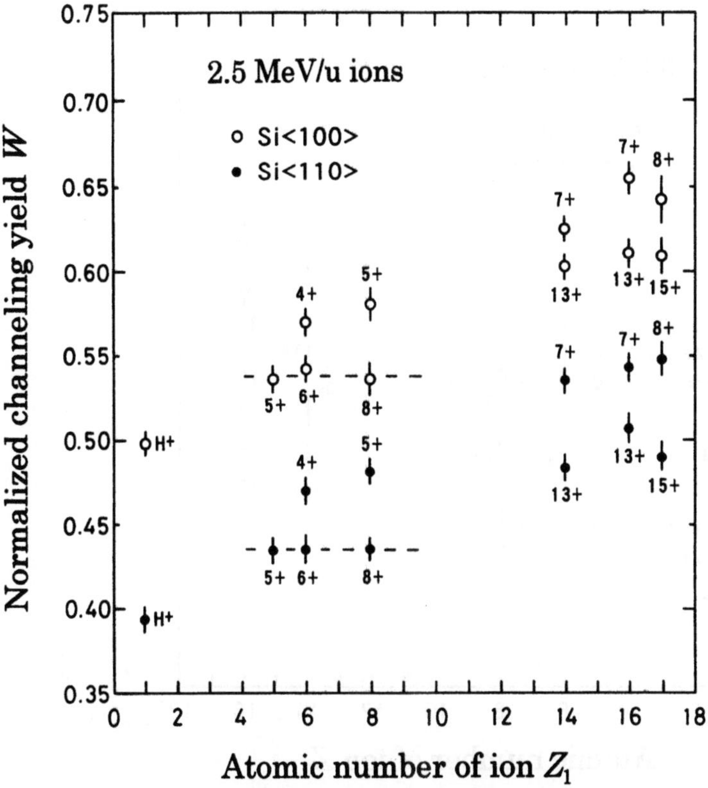

Fig. 9.4. Normalized channeling yield from Si at 3.3 keV, measured for various 2.5 MeV/u ions under ⟨100⟩ and ⟨110⟩ channeling incidence conditions. The input charge states of the ions are indicated near the plots. The *dashed lines* show the fully stripped levels [118]

and Si⟨110⟩, respectively, for 2.5 MeV/u B^{5+}, C^{6+}, and O^{8+}. This indicates that for any particular axial direction, the shadow cone radii R_c for these ions are the same. This certainly implies $Z_{\text{eff}} = Z_1$ in (9.1), so that the ions are still fully stripped in the crystal.

It can also be recognized that each value of W for H^+ is smaller than the fully stripped level. This is because the value of R_c for H^+ is greater than that for the fully stripped ions by a factor of $\sqrt{2}$, which leads to stronger shadowing of H^+ (Sect. 4.5.1). The values of W for the fully stripped ions and for protons are summarized in Table 9.1. The values of W for the other ions are greater than the fully stripped level, which indicates the existence of a screening effect due to captured inner-shell electrons.

The measured values of the normalized channeling yield were used to determine the values of μ, applying (8.1). The results are also shown in Table

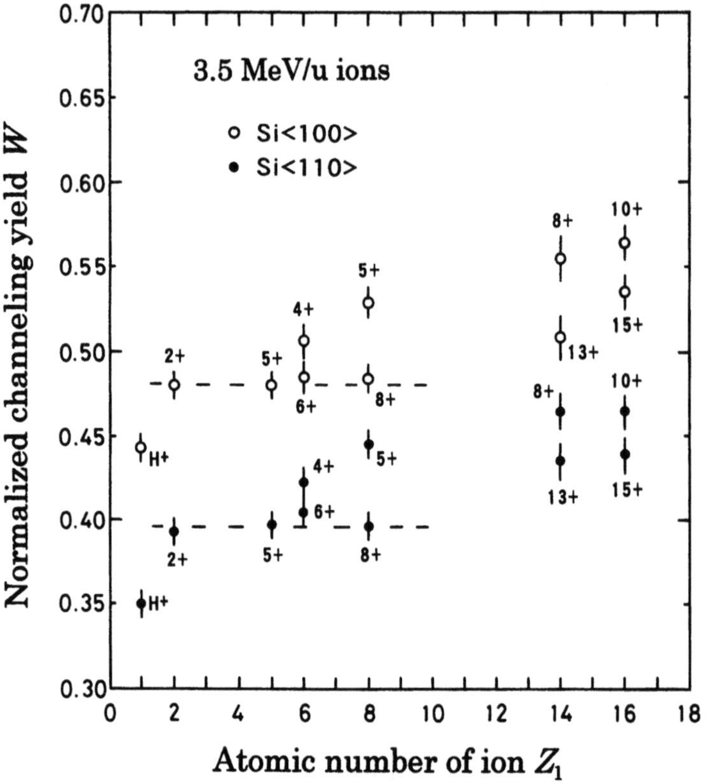

Fig. 9.5. Normalized channeling yield at 4.4 keV, measured for 3.5 MeV/u ions, shown similarly to Fig. 9.4. The *dashed lines* show the fully stripped levels [118]

9.1. An analysis of the values of μ obtained is not given here, since this subject has already been discussed in Sect. 8.5.

9.5.3 Charge States of Nonchanneled Ions

In the determination of Z_{eff}, the parameter q_r in (9.6) is assumed to be the most probable charge of the ions in the charge equilibrium, which is usually observed in carbon-foil transmission experiments. It has been assumed here that $q_r = 6$, 8, 12, 13.5, and 14 for the 2.5 MeV/u C, O, Si, S, and Cl ions, respectively, while $q_r = 6$, 8, 12.7, and 14 for the 3.5 MeV/u C, O, Si, and S ions, respectively [117]. This assumption seems reasonable for the higher input charges of the ions (C^{6+}, O^{8+}, Si^{13+}, S^{13+}, S^{15+}, and Cl^{15+}), since the ions should quickly reach charge equilibrium when the most probable charge is approximately equal to the input charge.

For the lower input charges of the ions (C^{4+}, O^{5+}, Si^{7+}, Si^{8+}, S^{7+}, S^{10+}, and Cl^{8+}), however, further investigation is required to test the validity of

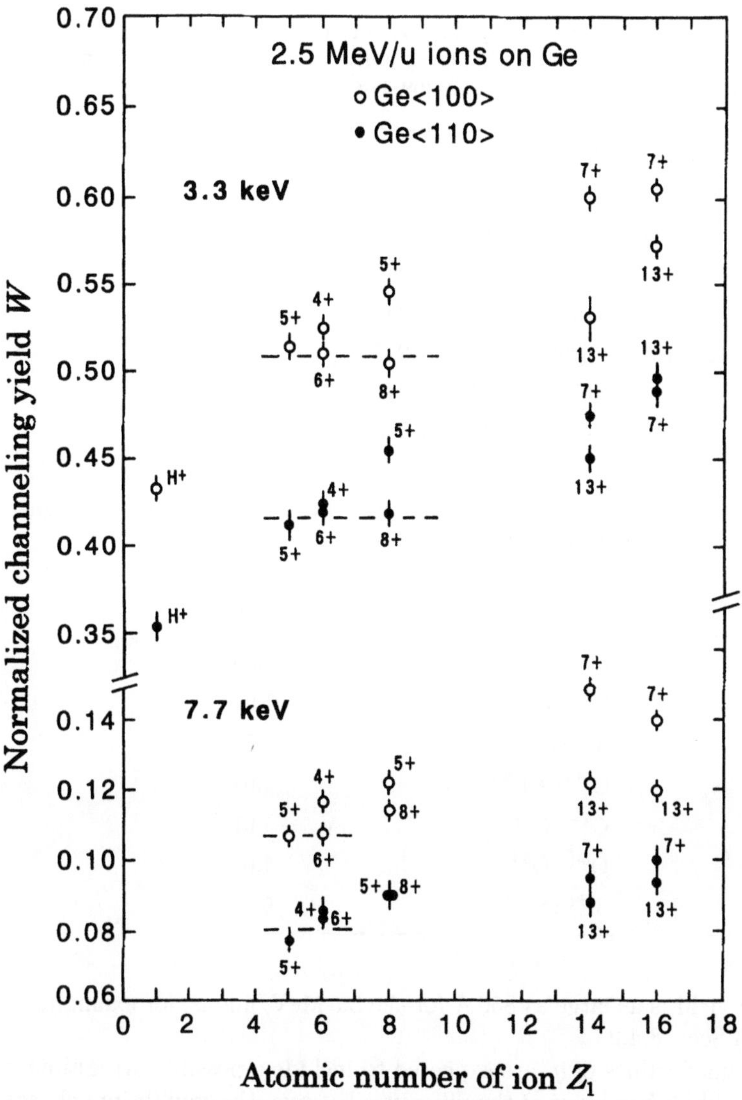

Fig. 9.6. Normalized channeling yield from Ge at 3.3 and 7.7 keV, measured for 2.5 MeV/u ions under ⟨100⟩ and ⟨110⟩ channeling incidence conditions. The *dashed lines* show the fully stripped levels [69]

the above assumption. Table 9.2 shows ratios of the random yield (per ion) for lower to higher input charges. The ratios are equal to unity within the estimated uncertainty of ~4%, from which we may conclude that in the random case the observed electrons are induced mainly by ions in equilibrium charge states. It is notable that the effective escape length responsible for

Table 9.1. Normalized channeling yields W_f for the fully stripped ions, W_p for protons, and the values of μ obtained. The estimated uncertainty in the normalized channeling yields is typically within ±3% [69, 118]

Ion energy (MeV/u)	Axis	W_f	W_p	μ
2.5	Si⟨100⟩	0.538	0.498	0.401
2.5	Si⟨110⟩	0.435	0.394	0.295
3.5	Si⟨100⟩	0.481	0.442	0.348
3.5	Si⟨110⟩	0.396	0.350	0.239
2.5	Ge⟨100⟩	0.509	0.432	0.246
2.5	Ge⟨110⟩	0.416	0.352	0.198

Table 9.2. Ratios of random electron yields at $\sim(E_\mathrm{B} + E_\mathrm{L})/2$ for lower to higher input charge states of ions incident on Si [118]

Ion energy (MeV/u)	Charge states (lower : higher)	Ratio
2.5	$C^{4+} : C^{6+}$	0.96
2.5	$O^{5+} : O^{8+}$	1.02
2.5	$S^{7+} : S^{13+}$	0.88
2.5	$Cl^{8+} : Cl^{15+}$	1.03
3.5	$C^{4+} : C^{6+}$	1.01
3.5	$O^{5+} : O^{8+}$	1.05
3.5	$Si^{8+} : Si^{13+}$	0.91

the random yield is as large as 850 Å for the 3.5 MeV/u ions, for example, as discussed in Sect. 8.4.2.

The assumed values of q_r are expected to include a possible overestimate of up to roughly 6% because of the difference between the equilibrium charge states for a carbon foil and for a Si crystal [117]. A decrease in q_r by 6% typically causes a decrease in the values of Z_eff by ~0.3 for the light ($Z_1 \leq 8$) ions and by ~0.5 for the heavy ions, which are only minor changes in the analysis.

9.5.4 Analysis Results

The results of the analysis are shown in Figs. 9.7–9.10, where the values of Z_eff obtained from (9.7) are plotted against Z_1. The numbers near the plots indicate the input charge states of the ions. The line $Z_\mathrm{eff} = Z_1 - 10$ shown in the figures, corresponding to Ne-like ions, indicates a lower limit on Z_eff since

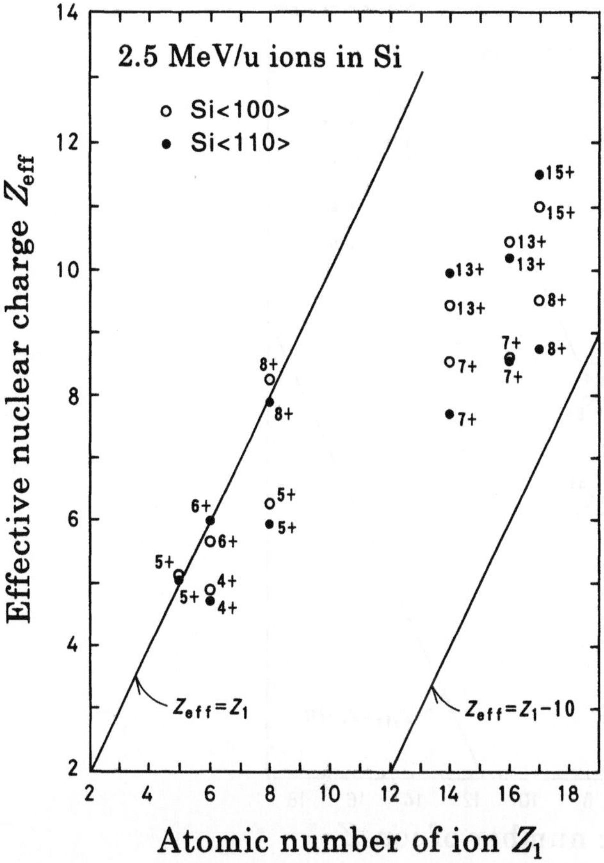

Fig. 9.7. Z_{eff} for the 2.5 MeV/u ions in Si plotted against Z_1. The input charge states of the ions are indicated near the plots. The line $Z_{\text{eff}} = Z_1$ corresponds to the fully stripped ions, while the line $Z_{\text{eff}} = Z_1 - 10$ represents a lower limit on Z_{eff} [118]

the ions should hardly capture electrons into the outer M shell while they are moving in the crystal. Indeed, the M-shell radii of the heavy ions are as large as ~1 Å, which is comparable to the size of the crystal channel so that there can be no bound M state. In the present analysis, it is hard to resolve any difference in Z_{eff}, neither between Si⟨100⟩ and Si⟨110⟩, nor between Ge⟨100⟩ and Ge⟨110⟩, and therefore it is better to discuss the averaged values of Z_{eff} for the two directions. Table 9.3 summarizes the input charges of the ions Z_{in}, the averaged values of Z_{eff} for the two directions Z^*_{eff}, and $\Delta Z = Z_{\text{in}} - Z^*_{\text{eff}}$ for the heavy ions. ΔZ corresponds to the number of electrons captured or lost ($\Delta Z > 0$ or $\Delta Z < 0$, respectively) in glancing collisions with the aligned atoms.

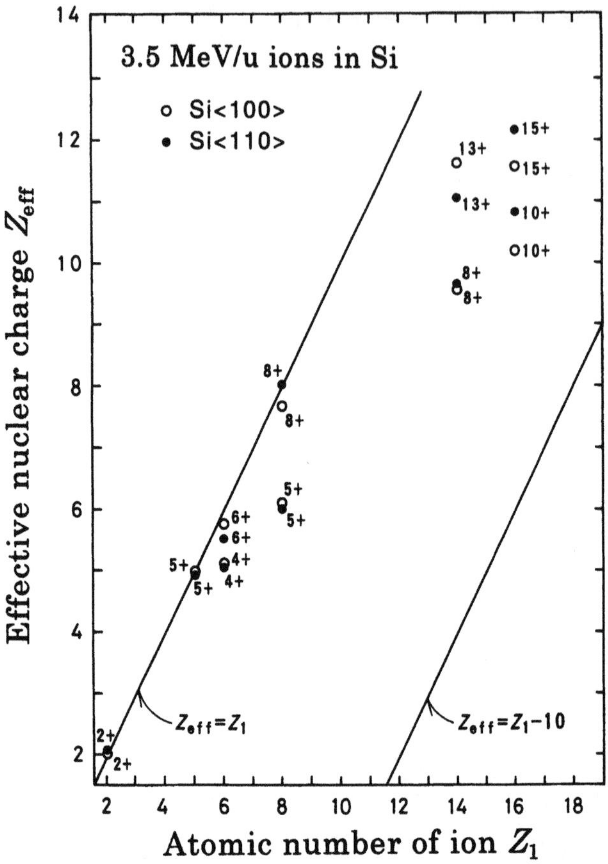

Fig. 9.8. Z_{eff} for the 3.5 MeV/u ions in Si plotted against Z_1, shown similarly to Fig. 9.7 [118]

We first focus on the charge states of relatively light ions, i.e. $Z_1 \leq 8$. For 2.5 MeV/u ions on a Si crystal, the values of Z_{eff} for B^{5+}, C^{6+}, and O^{8+} are on the line $Z_{\text{eff}} = Z_1$ which represents the fully stripped charge states of ions in the crystal, i.e. the *frozen* charge states. For C^{4+} and O^{5+}, the values of Z_{eff} are about 5 and 6, respectively (see also Table 9.3), indicating that each ion typically interacts with the aligned Si atoms after losing one electron. Similar results have been obtained for 3.5 MeV/u ions incident on a Si crystal. In this case, the charge states of the fully stripped He, B, C, and O ions are frozen, while C^{4+} and O^{5+} typically lose one electron prior to the glancing collisions with Si atoms. The charge states are very similar for a Ge crystal.

In contrast, electron capture as well as loss occurs for Si, S, and Cl ions in Si and Ge crystals. In a Si crystal, 2.5 MeV/u Si^{13+}, S^{13+}, and Cl^{15+} capture about three or four electrons, whereas Si^{7+}, S^{7+}, and Cl^{8+} lose one

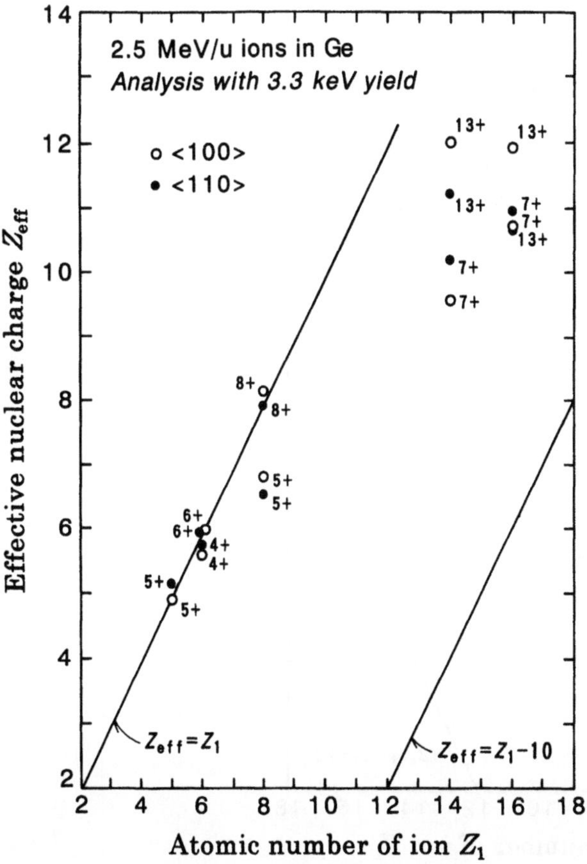

Fig. 9.9. $Z_{\rm eff}$ for the 2.5 MeV/u ions in Ge plotted against Z_1, obtained from the 3.3 keV yield, shown similarly to Fig. 9.7 [69]

or two electrons, as seen in Table 9.3. Therefore, we may conclude that in the channeling case the most probable charges in equilibrium, $Z_{\rm eq}$, in a Si crystal are $8.2 < Z_{\rm eq} < 9.7$, $8.6 < Z_{\rm eq} < 10.3$, and $9.1 < Z_{\rm eq} < 11.3$ for 2.5 MeV/u Si, S, and Cl, respectively. Similarly, it is concluded that in a Si crystal $9.6 < Z_{\rm eq} < 11.3$ and $10.5 < Z_{\rm eq} < 11.9$ for 3.5 MeV/u Si and S, respectively. In the channeling case the number of electrons bound by the ions is greater than in the random case (see Sect. 9.5.3) by 2–4, 3–5, and 3–5 for 2.5 MeV/u Si, S, and Cl, respectively, and by 1–3 and 2–4 for 3.5 MeV/u Si and S, respectively. It can also be recognized that $Z_{\rm eq}$ for Si and S ions increases with increasing ion velocity. For a Ge crystal, each value of $Z^*_{\rm eff}$ is greater than that for a Si crystal by typically 1–2, indicating that the electron loss is enhanced relative to the case of Si. Since the results obtained from the 3.3 and 7.7 keV electron yields are essentially the same, the analysis results for the 3.3 keV case only are presented in Table 9.3. The values of $Z_{\rm eq}$ for a Ge

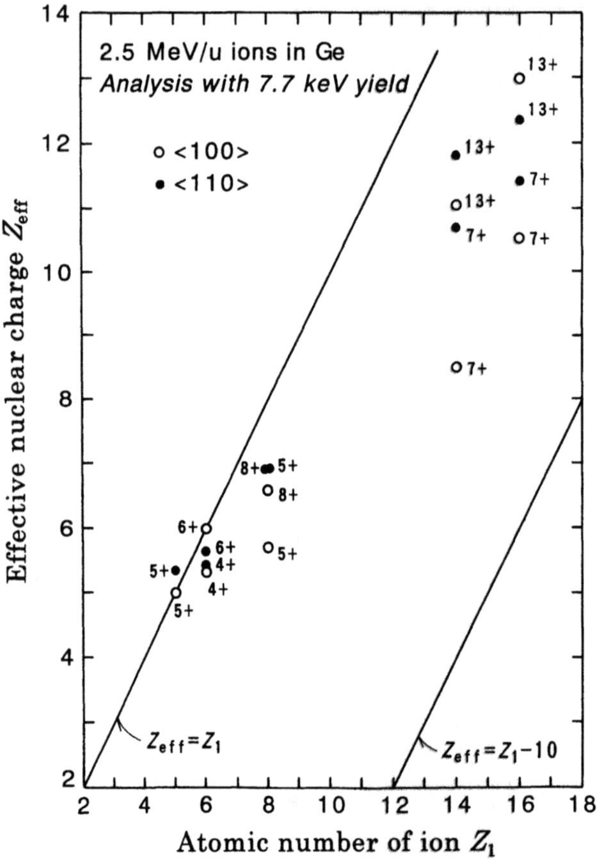

Fig. 9.10. Z_{eff} for the 2.5 MeV/u ions in Ge plotted against Z_1, obtained from the 7.7 keV yield, shown similarly to Fig. 9.7 [69]

crystal can be also estimated from the 3.3 keV electron yield, for example, as $9.9 < Z_{eq} < 11.6$ and $10.8 < Z_{eq} < 11.3$ for 2.5 MeV/u Si and S, respectively. These values are greater than those in a Si crystal by 1–2, corresponding to the enhanced values of Z_{eq} noted above.

9.5.5 Two Competitive Screening Effects

It can be recognized that the increase in the normalized channeling yield W for the partially stripped ions results roughly equally from the reduced shadowing effect and the enhanced emission effect. For example, for 2.5 MeV/u Cl^{15+} incident on Si⟨100⟩ (for which $Z_{eff} = 11$ has been determined), the enhanced emission effect increases the value of W by a factor of $F(17,11)/F(17,14) = 1.07$.[1] Therefore, the enhanced emission effect is

[1] In this case, $\xi(Z_1) = 0.41$ for 1.36 keV electrons.

Table 9.3. Z_{in}, Z^*_{eff} (averaged value of Z_{eff} for ⟨100⟩ and ⟨110⟩), and $\Delta Z = Z_{\text{in}} - Z^*_{\text{eff}}$, determined for Si and Ge crystals [69, 118]

Ion energy (MeV/u)	Ion	Z_{in}	Z^*_{eff} in Si	Z^*_{eff} in Ge	ΔZ in Si	ΔZ in Ge
2.5	C	4	4.8	5.7	−0.8	−1.7
2.5	O	5	6.1	6.7	−1.1	−1.7
2.5	Si	13	9.7	11.6	3.3	1.4
2.5	Si	7	8.2	9.9	−1.2	−2.9
2.5	S	13	10.3	11.3	2.7	1.7
2.5	S	7	8.6	10.8	−1.6	−3.8
2.5	Cl	15	11.3		3.7	
2.5	Cl	8	9.1		−1.1	
3.5	C	4	5.1		−1.1	
3.5	O	5	6.0		−1.0	
3.5	Si	13	11.3		1.7	
3.5	Si	8	9.6		−1.6	
3.5	S	15	11.9		3.1	
3.5	S	10	10.5		−0.5	

responsible for about 50% of the enhancement factor of 0.609/0.538 = 1.13 for the observed value of W relative to the fully stripped case (see Fig. 9.4).

To illustrate this in more detail, Fig. 9.11 illustrates the determination of Z_{eff} for the ⟨100⟩ incidence of 2.5 MeV/u Si^{7+} on Ge. Z_{eff} can be determined from the calculated Z_{eff} dependence of W, shown by the solid curve, and the observed value of $W_s = 0.598$ for the 3.3 keV electron yield. It should be noted that the solid curve gives a normalized yield of 0.484 at $Z_{\text{eff}} = 14$, which is lower than W_f for fully stripped light ions, because of the enhanced random yield corresponding to $q_r = 12$ (instead of 14) for the Si ions, noted in Sect. 9.5.3. The dashed curves in Fig. 9.11 show the calculated Z_{eff} dependence of W assuming the reduced shadowing effect only or the enhanced emission effect only. The calculations considering only the former or only the latter effect conclude that $Z_{\text{eff}} \simeq 8$ and $Z_{\text{eff}} \simeq 5$, respectively, as indicated by Z_α and Z_β on the abscissa in Fig. 9.11. These are underestimates by about 2 and 5, respectively, compared with the determined value of $Z_{\text{eff}} = 9.56$. This demonstrates that the two screening effects make similar contributions to the enhancement of the value of W.

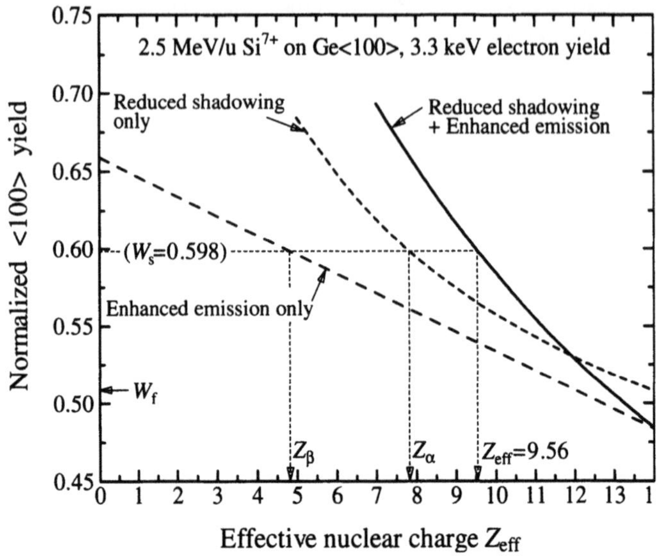

Fig. 9.11. Determination of $Z_{\rm eff}$ for $\langle 100\rangle$ incidence of 2.5 MeV/u Si^{7+} on Ge from the value of $W_s = 0.598$ measured for the 3.3 keV electrons. The calculated dependence of the ratio on $Z_{\rm eff}$, shown by the *solid curve*, indicates $Z_{\rm eff} = 9.56$. The *dashed curves* show the ratio calculated assuming the reduced shadowing effect only or the enhanced emission effect only. $W_f = 0.509$ indicates the value for fully stripped ions [69]

9.5.6 Nonequilibrium Charge States

It is of fundamental importance to estimate the thickness of the surface layer that is required to establish the charge states determined in the present analysis. For this purpose, we estimate the effective surface layer thickness t beyond which most of the ions are deflected away from the rows of atoms and contribute less to the electron production. The charge states determined by the present analyses should be those for ions penetrating from the suface to a depth of t. The value of t can be estimated either from experiments using overlaid crystals or from numerical calculations, as shown in Sect. 8.4. It is of course better to use experimental values since the calculations provide only approximate values. For the electron yield at 3.8–4.8 keV induced by 3.5 MeV/u ions, we have already determined $t \simeq 175$ Å for Si$\langle 100\rangle$ in Sect. 8.4.2. The values of t for other cases can be estimated using the trajectory scaling with respect to d or V_1, as discussed in Sect. 4.5.1. It follows that $t \simeq 147$ Å for the 3.5 MeV/u ions on Si$\langle 110\rangle$, and, for the 2.5 MeV/u ions, $t \simeq 148$ and 124 Å for Si$\langle 100\rangle$ and Si$\langle 110\rangle$, respectively. Note that in these estimates a possible difference between the values of t in the screened and unscreened cases has been ignored, and that the difference in the impact pa-

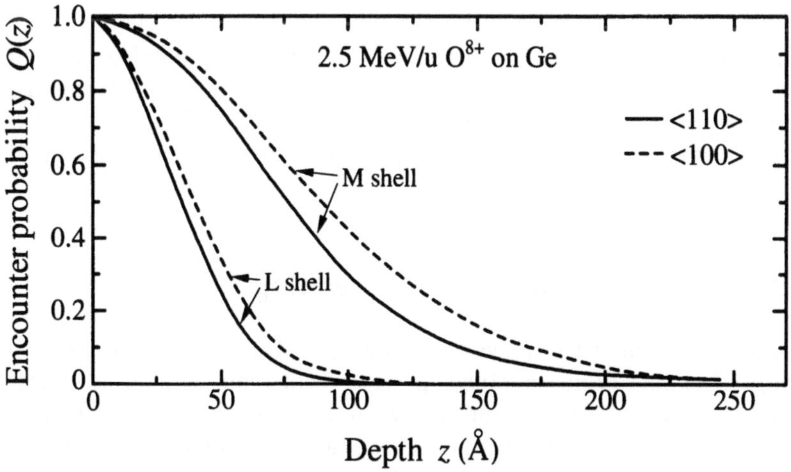

Fig. 9.12. Calculated encounter probabilities for the L- and M-shell electrons under ⟨110⟩ and ⟨100⟩ channeling incidence conditions of 2.5 MeV/u O^{8+} on Ge, normalized to the value at the surface [69]

rameter dependence of the binary-encounter production at 2.5 and 3.5 MeV/u has also been neglected.

We may conclude from the above values of t that the charge states determined in this manner are established well before the ions have suffered 30–40 glancing collisions with the aligned Si atoms in the ⟨100⟩ or ⟨110⟩ direction. Figure 9.12 shows $Q(z)$ for Ge L- and M-shell electrons calculated under ⟨110⟩ and ⟨100⟩ channeling incidence conditions of 2.5 MeV/u O^{8+} on Ge, normalized to the value at the surface. This indicates that both the L and M shells are well shadowed at a penetration distance of about 150 Å, which corresponds to roughly 30 times the interatomic distance. It may be concluded that the charge states determined for Ge crystals are, on the average, those of ions that have experienced 10–20 collisions with the aligned Ge atoms, similarly to the case for Si. We also see from Fig. 9.12 that the number of collisions needed to shadow the Ge M shell is roughly twice as many as for the L shell. The charge states determined from the 7.7 keV yield should have a greater contribution from the inner L shell than from the outer M shell compared with the charge states determined from the 3.3 keV yield, as is anticipated from the electron production cross sections calculated by the binary-encounter model. As can be seen from Figs. 9.9 and 9.10, however, no significant difference between the values of Z^*_{eff} determined from the 3.3 and 7.7 keV yields has been found in the present analysis. This probably indicates that the main contribution to the electron yield at the two electron energies comes from the Ge M shell.

It must be emphasized that in the channeling case the observed electron yield is produced mainly by ions running close to the atomic rows, where many electrons should be recoiled by the ions. Therefore, the charge states determined by the present method are characteristic of ions passing through a region of a high density of electrons near the atomic rows and, accordingly, they must be different from those observed by transmission channeling experiments, in which the charge state of ions encountering a low density of electrons is measured.

9.6 Effective Reduction of the Screening Length

In the model used here to analyze the charge states of ions, the reduced shadowing effect has been explained in terms of a reduced nuclear charge of the ion. It is of fundamental interest to investigate whether the reduced shadowing effect for partially stripped ions can be interpreted as a reasonable reduction of the screening length for the most commonly used interaction potential. This is essentially an experimental check of an approximate expression for the ion–atom interaction potential in terms of the Firsov screening length given by (4.30). For this purpose, an approach in terms of the screening length for the Molière potential has been followed.

In the Molière potential, the screening length for a fully stripped ion a is given by (4.29), which depends on Z_2 only. For a pair of atoms, the Molière potential is still a good approximation if a is simply replaced by (4.30), which depends on both Z_2 and Z_1. The channeled trajectories are mainly determined by collisions with impact parameters of the order of 0.1 Å. Therefore, the high-energy shadowing effect of ions that have filled inner shells but no electrons in the outer shells must be the same as that of neutral projectiles and, consequently, their screening lengths should be equal.

To estimate the dependence of W on the screening length, computer simulations of the multistring type described in Sect. 4.6 have been carried out. It was assumed that the impact-parameter dependence of the probability of close-encounter collisions is proportional to the two-dimensional electron density of the target atoms. As seen in Sect. 8.4, such calculations tend to underestimate the effective surface layer thickness for production of binary-encounter electrons. Nevertheless, we may anticipate that typical aspects of the screening effect can be deduced from the numerical approach.

Figure 9.13 shows the encounter probability for Si ⟨110⟩ calculated for close-encounter collisions of 3.75 MeV/u ^{16}O ions with $L_{2,3}$ shell electrons in Si for reduced screening lengths of $0.4a$ and $0.6a$ as well as a. Clearly, the shadowing effect reduces with decreasing screening length, i.e. with reduction of the repulsive force exerted on the ions.

The effective surface layer thicknesses were obtained by integrating the probabilities from the surface to a depth of ~400 Å, where the first glancing collisions are completed, as discussed in Sect. 8.4.3. The results are shown in Fig. 9.14 as a function of the screening length, for K, L_1, and $L_{2,3}$ shells in Si and 3.75 and 1.875 MeV/u ions under ⟨110⟩ and ⟨100⟩ channeling incidence

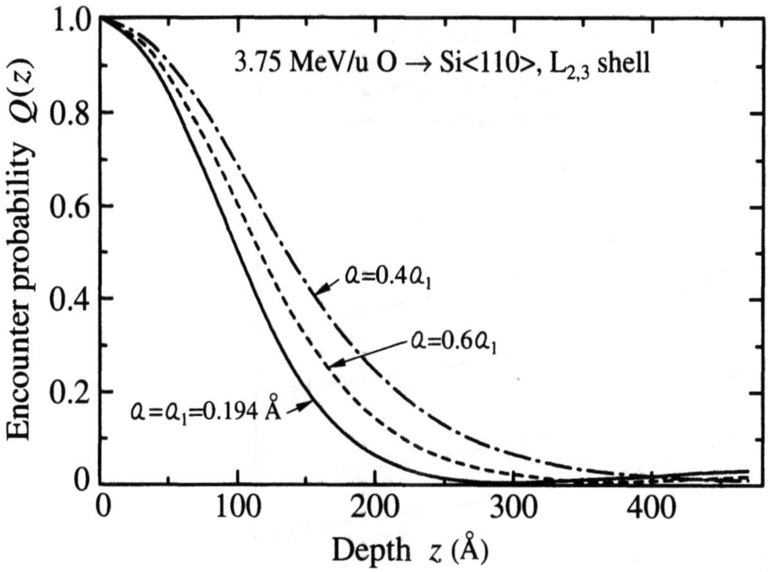

Fig. 9.13. Dependence of the encounter probability for Si $L_{2,3}$-shell electrons on the screening length of the Molière potential, calculated for ⟨110⟩ incidence of 3.75 MeV/u O ions on Si at room temperature [124]

conditions on Si, as well as for the L ($L_1 + L_{2,3}$) shell in GaAs in the ⟨100⟩ case. For Si atoms, the values of the Firsov screening length are 0.122, 0.120, and 0.118 Å for Si, S, and Cl ions, respectively; these values are less than the Thomas–Fermi screening length (0.194 Å) by a factor of ~0.6 (Fig. 9.13). As seen in Fig. 9.14, this difference in the screening length gives rise to an enhancement of the effective surface layer thickness for both the ⟨110⟩ and the ⟨100⟩ case in Si, typically by factors of 1.10 and 1.15 for K- and L-shell electrons, respectively, relative to the values for the Thomas–Fermi screening length. A similar enhancement (by a factor of ~1.1) is seen for the L shell in GaAs, for which the Firsov screening length is shorter than the Thomas–Fermi screening length (0.147 Å) by a factor of ~0.75. The effective surface layer thickness for the L shell is more sensitive to the screening length than that for the K shell because the trajectories of ions for different values of the screening length diverge from each other only at large distances from the atomic rows.

In view of the derivation of (9.7), the enhancement factor noted above should be approximately equal to

$$\frac{W_{\mathrm{mod}} - \mu}{W_{\mathrm{f}} - \mu} = \sqrt{\frac{Z_1}{Z_{\mathrm{eff}}}}\,. \tag{9.8}$$

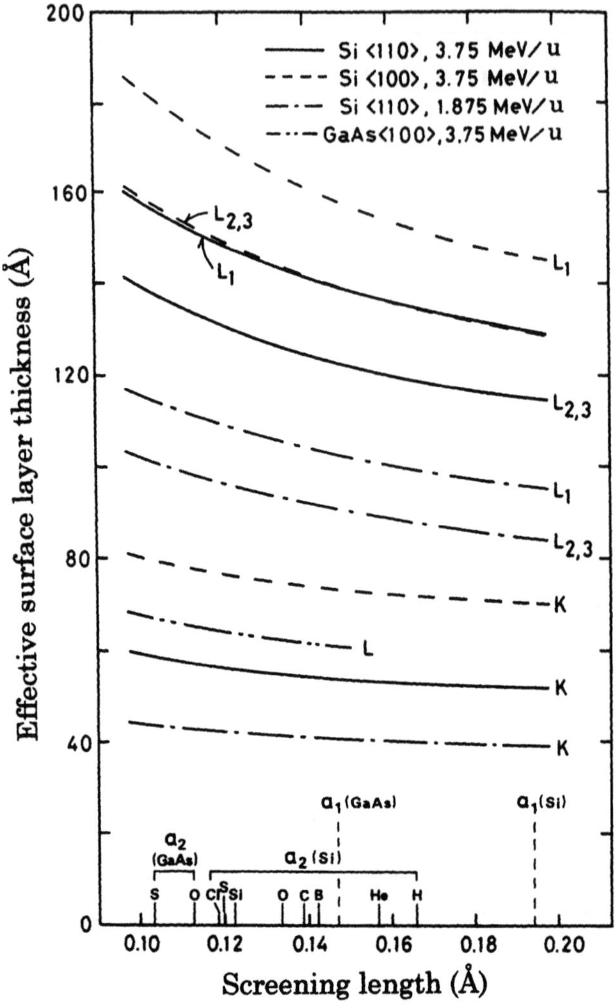

Fig. 9.14. Effective surface layer thickness for electronic shells in Si and GaAs as a function of the screening length of the Molière potential, calculated for ⟨110⟩ and ⟨100⟩ channeling incidence of 1.875 and 3.75 MeV/u ions; a_1 and a_2 indicate the screening lengths given by (4.29) and (4.30), respectively [124]

In the present analysis, Z_{eff} in (9.8) was replaced by the averaged value Z^*_{eff}, shown in Table 9.3. Figure 9.15 shows plots of $\sqrt{Z_1/Z^*_{\mathrm{eff}}}$ against Z_1 for the various values of Z^*_{eff}. Most of the points are roughly in the vertical range 1.1–1.2, in good agreement with the calculated enhancement factors,[2] although

[2] In an earlier analysis [118], the enhanced emission effect of partially stripped ions was not taken into account, which led to an overestimate of the experimental enhancement factors.

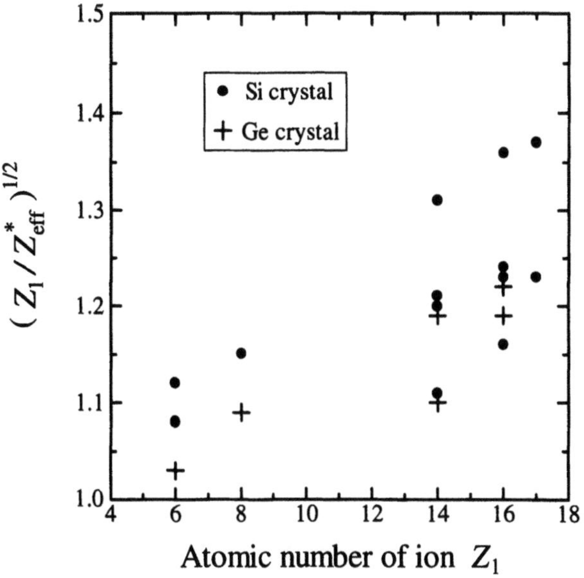

Fig. 9.15. Plots of $\sqrt{Z_1/Z^*_{\text{eff}}}$ against Z_1

some values are greater than 1.3, corresponding to 2.5 MeV/u Si^{7+}, S^{7+}, and Cl^{8+} incident on a Si crystal; these are underestimated by a factor of 0.8–0.9 by the present estimate, which uses the Firsov screening length.

From the analysis given here, we may conclude that the observed increase in the electron yield under channeling incidence conditions of heavy ions as compared with light ions can be interpreted in terms of a reduction of the screening length in the Molière potential. The analysis has shown the applicability of the Firsov screening length to partially stripped ions. It is useful to compare the screening for Si, S, and Cl ions with that for O ions since they have nearly equal values of the Firsov screening length. However, for O^{8+} ions the Thomas–Fermi screening length must be used since these ions are fully stripped in the crystals considered here. In contrast, the heavy ions with the same velocity have several electrons and, although those ions are not neutral, the analysis here suggests that the Firsov screening length (for neutral projectiles) is applicable. The presence or absence of outer-shell electrons is unimportant in the interaction between an ion and the atoms of the crystal, as mentioned earlier in this section.

10. Materials Analysis with Binary-Encounter Electrons

Binary-encounter electron spectroscopy (BEES) in conjunction with high-energy shadowing can be applied to the structural analysis of crystalline materials. The principle of analysis is similar to that well established in ion backscattering spectroscopy. The characteristic aspects of this method and several examples of applications are presented in this chapter.

10.1 BEES and Ion Backscattering Analysis

BEES is an analysis technique for crystalline materials, which is complementary to Rutherford backscattering spectroscopy (RBS) in particular. While RBS analysis is essentially based on the detection of nuclei in the crystal, BEES reflects the spatial distribution of the target electrons, as seen in the preceding chapters. In this sense, structural information obtained from BEES is rather equivalent to that from X-ray diffraction. The characteristic aspects of BEES can be summarized as follows:

- practical use of high-energy ions
- shadowing effect on the most populated atomic shell
- applicability of ions heavier than the target atoms
- low beam-current or low-beam-dose analysis.

In BEES we may use ions of higher energy than those commonly used in ion backscattering analysis, even when non-Rutherford, i.e. nuclear elastic, scattering of the ions is considerable. This enables high-energy shadowing measurements under narrow-critical-angle conditions, which allow precise analysis of distorted crystals, for example. The binary-encounter electron yield of interest is typically in the keV energy range for incident ions in the MeV/u energy range. The keV electron yield is sensitive to the effect of high-energy shadowing on the most populated electronic shell of the target atoms, such as the L shell for Si or the M shell for Ge.

It is notable that heavy ions can be used to produce binary-encounter electrons without any limitations imposed by the atom species of the target material, in contrast to ion backscattering spectroscopy, which requires ion species lighter than the target atoms. Furthermore, the high count rate of the electrons enables analysis with a beam current or beam dose that is lower by a factor of typically 10^2 to 10^3 than that required for ion backscattering spectroscopy. This also allows a quick measurement of the shadowing pattern for a spot area of the crystal surface, which is useful for structural characterization of crystalline materials.

BEES is also applicable to insulators under the condition that most of the ions penetrate the sample and reach the electrically-conducting substrate, so that no charging effect occurs (Sect. 5.2). In many cases, such conditions can be satisfied by an appropriate choice of the incident ion energy so that the mean penetration length of the ions exceeds the thickness of the sample.

On the other hand, there are demerits compared with ion backscattering spectroscopy. The typical depth sensitivity in ion-induced-electron analysis is several to one hundred nanometers and, accordingly, is less suitable for surface structure analysis. Of course, this is due to the fact that the binary-encounter electron yield reflects high-energy shadowing in the surface layer up to a depth equal to the effective escape length of the electrons (Sect. 6.2). Furthermore, the effective emission cross section for the electrons from a solid target cannot be easily predicted, unlike the case for Rutherford backscattering of the ions. For wider use of BEES, a database is needed that will provide a quick and reliable estimate of the effective escape length, which represents the depth sensitivity of the analysis.

The following sections demonstrate several applications of BEES to materials analysis which rely on the technical merits noted above.

10.2 Misoriented Crystal Lattice

BEES can be applied to the structural characterization of epitaxially grown crystals [137–140]. In the following, an analysis of CeO_2 crystal films of ~1000 Å thickness is described. CeO_2 has the fluorite structure and can be grown epitaxially on Si [141, 142]. CeO_2 samples prepared under various growth conditions were studied by BEES with 56 MeV (3.5 MeV/u) O^{8+} and also by RBS with 1.5 MeV He^+. In the latter, the channeling minimum yield χ_{min} was measured to investigate the lattice defects. Figure 10.1 shows typical energy spectra of secondary electrons measured at a backward angle of 180° under ⟨110⟩ channeling and random incidence conditions, together with the normalized channeling yield. In the measurements, the angle between the ⟨110⟩ direction used and the surface normal of the epitaxially grown $CeO_2(111)$ film on a Si(111) substrate was ~35°. For characterization of the CeO_2 samples, the normalized electron yield W for the ⟨111⟩ axis normal to the surface was measured for an energy window of 7.7–9.0 keV, which is slightly above $E_B = 7.6$ keV.

Typical RBS spectra for Ce in a CeO_2 layer of 3000 Å thickness, measured at a backward angle of 150°, are shown in the inset in Fig. 10.2. The χ_{min} values were determined at a backscattered energy of 1.32 MeV, which is on the

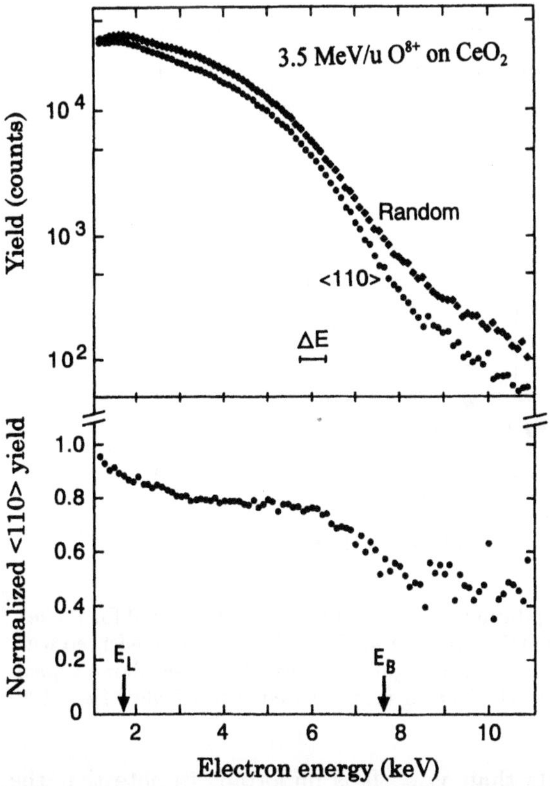

Fig. 10.1. Energy spectra of secondary electrons induced by 3.5 MeV/u O^{8+} from CeO_2 under ⟨110⟩ and random incidence conditions, and the normalized ⟨110⟩ yield obtained from the spectra [138]

low-energy side of the surface peak seen in the ⟨111⟩ spectrum. In this case, the estimated thicknesses of the surface layer responsible for χ_{min} and W are both roughly 200 Å.[1] It should be noted that the commonly used parameter χ_{min} was determined with the surface peak yield excluded. In contrast, W necessarily includes the surface peak yield since the electron yield stems from all the unshadowed atoms in a surface layer whose thickness is equal to the effective escape length. This difference accounts qualitatively for the fact that the observed values of W are greater than those of χ_{min}.

Lattice defects in the surface layer enhance the values of W and χ_{min} by direct scattering and dechanneling processes (Sect. 4.7). A χ_{min} versus W plot obtained for the CeO_2 samples is shown in Fig. 10.2. In the range W = 0.48–0.60, χ_{min} remains constant (~0.04), indicating that W is more

[1] The effective surface layer thickness for the electron yield was estimated from the calculated effective surface layer thickness (mainly for the M and N shells of Ce) and the observed value of W, using (8.2).

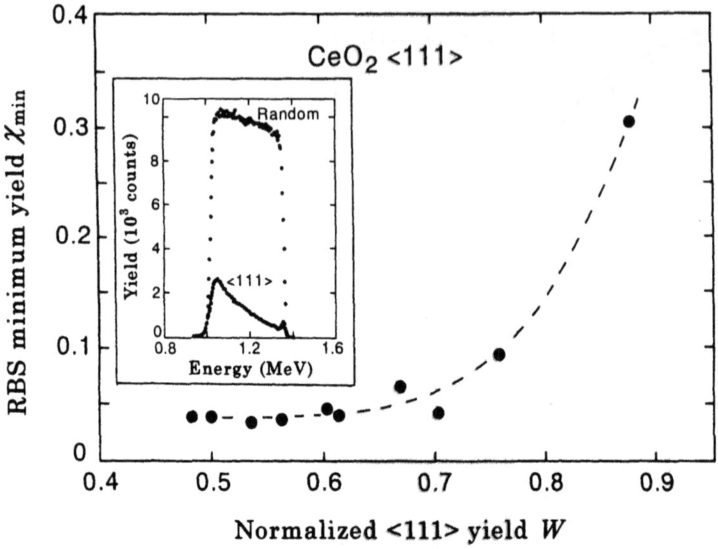

Fig. 10.2. χ_{min} vs. W plot for epitaxially grown CeO_2. W was measured for 7.7 keV electrons induced by 3.5 MeV/u O^{8+} ($E_B = 7.6$ keV), while χ_{min} was measured with 1.5 MeV He^+. The *dashed curve* is drawn to guide the eye. The *inset* shows typical RBS spectra for Ce in a 3000 Å thick CeO_2 layer, measured with 1.5 MeV He^+ [139]

sensitive to the lattice defects than χ_{min}. It is important to note that the channeling critical angle of ~0.3° for 56 MeV O^{8+} is narrower than that for 1.5 MeV He^+ by a factor ~0.3. The higher sensitivity to lattice defects for a narrower critical angle implies that the crystal contains a misorientation of the atomic rows which is of the order of the critical angle. The observed misorientation can be attributed to a mosaic structure or to extended defects such as dislocations. The thickness of the layer analyzed can be varied by varying the ion energy and the measured electron energy. Similar χ_{min} versus W relations have been obtained for imperfect Ni crystals [69].

RBS becomes less practical as the ion energy increases toward the non-Rutherford region (for example, for 3.5 MeV/u O^{8+} on Si) because of the complicated backscattering cross sections. Nevertheless, such high-energy ion beams can be still used for material analysis by BEES under narrow-critical-angle conditions. Electron measurements with a narrower critical angle for channeling should allow higher-sensitivity analysis of a misoriented crystal lattice.

10.3 Polarity of the Zinc Blende Structure

The zinc blende lattice structure has crystallographic polarity so that the configuration of atoms is not equivalent when it is seen along the [111] and

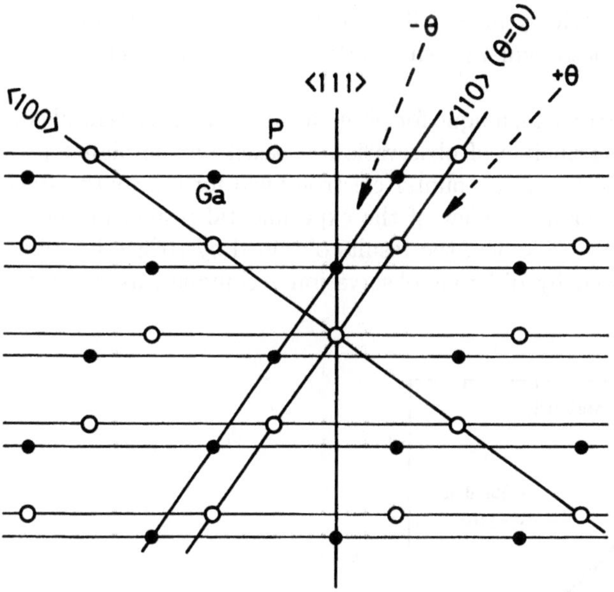

Fig. 10.3. Atomic arrangement of atoms on GaP(110), illustrating the preferential collision of ions with Ga ($\theta < 0$) or P ($\theta > 0$) atoms for incidence angles slightly off $\langle 110 \rangle$ ($\theta = 0$) [143]

$[\bar{1}\bar{1}\bar{1}]$ directions, as shown in Fig. 10.3. The determination of the polarity of zinc blende crystals has been of both fundamental and technical interest in the structural analysis of typical III–V and II–VI compound materials.

Ion beam analysis using MeV ions has become a powerful technique for the studies of this problem since Bontemps et al. [144, 145] first reported an asymmetric $\langle 110 \rangle$ channeling dip of the ion backscattering yield from GaP when the beam is scanned angularly along the (110) plane. The ion beam technique is particularly useful for the polarity determination of submicron-thick crystals on substrates. The well-known X-ray absorption method is less suitable for such thin crystals because of the long escape length (much longer than 10 μm) of diffracted X-rays relative to the crystal thickness.

The asymmetric $\langle 110 \rangle$ channeling dip results from the preferential collision geometry realized at grazing angles of incidence $\pm\theta$ with respect to $\langle 110 \rangle$, as shown schematically in Fig. 10.3. For the $-\theta$ incidence the predominant fraction (75%) of the beam first collides with the Ga rows, while the minor 25% fraction collides with the P rows. In this case, the ion backscattering yield from Ga is higher than from P. The preferential collision effect is reversed for $+\theta$. It has been confirmed that preferential collision occurs only near the crystal surface, where single collision of the ions with the atomic rows is dominant [146–149]. The preferential collision effect has been applied not only to polarity determination [150], but also to the study of long-range order in thin $(GaSb)_{1-x}(Ge_2)_x$ films [148] and the lattice site location of Te in GaAs [149]. Similar measurements using particle-induced X-rays have been also reported [151].

BEES provides a further technique for observation of the preferential collision effect [143]. The principle of observation is based on the higher production efficiency of the binary-encounter electron yield for a heavier target atom. This seems reasonable in view of the experimental results presented in Sect. 8.4.4. To confirm this idea, the results obtained by BEES described below were cross-checked by different observation techniques, as discussed later.

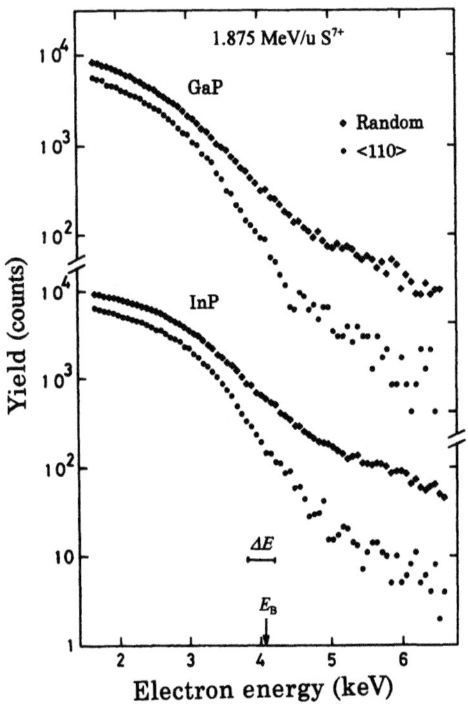

Fig. 10.4. Energy spectra of secondary electrons emitted from GaP and InP when bombarded by 1.875 MeV/u S^{7+} under ⟨110⟩ and random incidence conditions [143]

Figure 10.4 shows energy spectra of secondary electrons induced by 1.875 MeV/u S^{7+} under ⟨110⟩ and random incidence conditions on GaP and InP. The normalized ⟨110⟩ yield decreases rapidly at $\sim E_B = 4.1$ keV with increase of electron energy, indicating that above E_B the effect of shadowing on the inner-shell electrons is dominant. The angular dependence of the electron yield was measured for two energy ranges, below and above E_B, namely 2.6–2.8 and 5.5-6.7 keV. Figures 10.5 and 10.6 show the results for GaP and

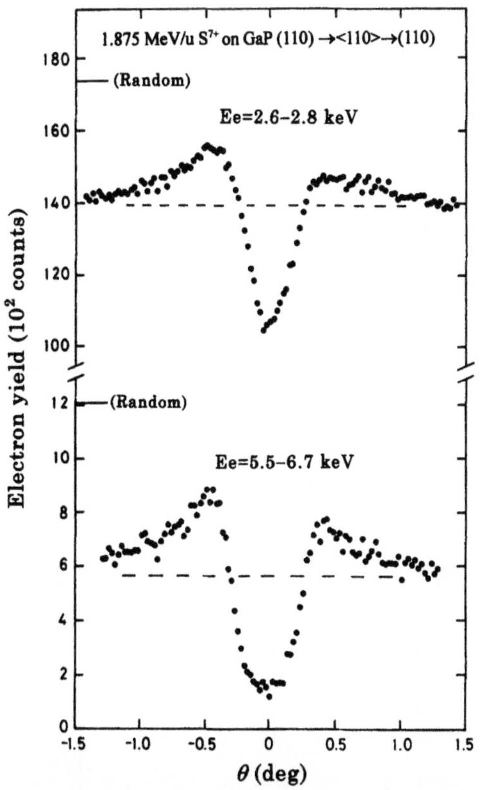

Fig. 10.5. Angular dependence of the electron yield across the ⟨110⟩ direction ($\theta = 0$) of GaP, scanned along the (110) planar direction. The *dashed lines* show the (110) yields. The yields for the random case are indicated [143]

InP, respectively, for an angular scan across ⟨110⟩ along the (110) planar direction, as shown in Fig. 10.3. Clearly, the angular dips are asymmetric with respect to the ⟨110⟩ direction ($\theta = 0$). The higher electron yield at $\theta \simeq -0.5°$ than at $\theta \simeq 0.5°$ results from the enhanced number of recoiled electrons due to the preferential collision of the ions with the heavier Ga or In atoms in the target crystals. Figure 10.7 shows RBS spectra of 1.875 MeV/u S^{7+} measured on InP for a cross-check of the BEES results. Note that the S ions are backscattered from the heavy In atoms only. The backscattering yield at ~25 MeV is enhanced at $\theta = -0.5°$ relative to the yield at $\theta = 0.5°$, which is consistent with the BEES results. Furthermore, the BEES results for GaP were consistent with a preanalysis of the (111) and $(\bar{1}\bar{1}\bar{1})$ surfaces by chemical etching [152].

It is of further interest to compare the ⟨110⟩ angular dips at energies above and below E_B. In Figs. 10.5 and 10.6, the dip above E_B is wider than

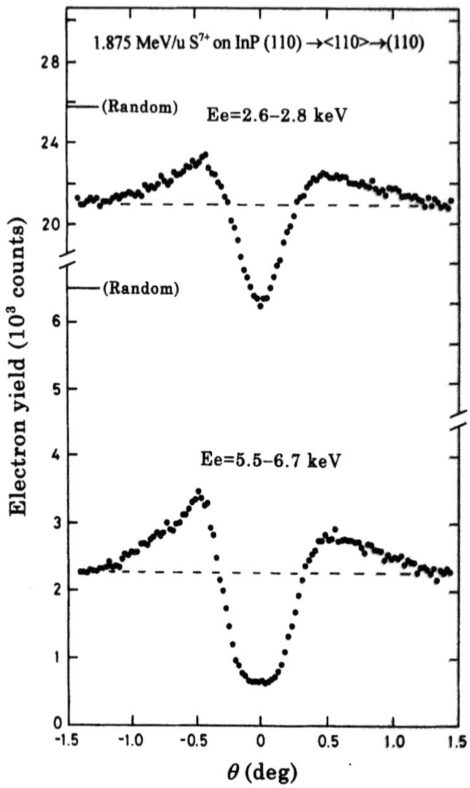

Fig. 10.6. Angular dependence of the electron yield for InP, shown similarly to Fig. 10.5 [143]

below E_B, demonstrating the selective observation of inner-shell shadowing above E_B.

10.4 Backward Channeling Microscopy

Historically, the pioneering work on channeling microscopy was reported by a reseach group in France [153]. They successfully observed transmission channeling images of lattice defects in thin crystals, using radioisotopes as an α-ray source. The key technique is based on the difference in the energy losses suffered by the channeled and dechanneled ions penetrating the crystal. A more refined version of transmission channeling microscopy has been developed recently with improvement of ion sources and beam transport systems of ion accelerators. The spatial resolution of transmission channeling microscopy using microbeams is enough to resolve extended defects such as stacking faults [154]. When channeling microscopy is applied to surface layer

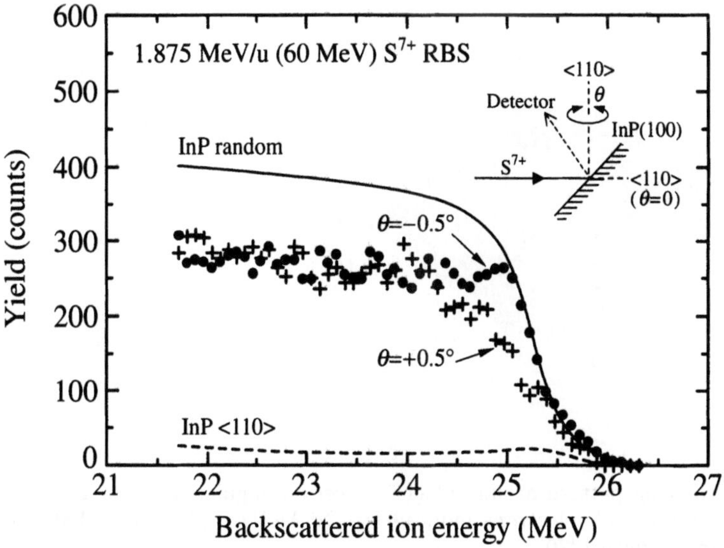

Fig. 10.7. RBS spectra for InP at $\theta = \pm 0.5°$ from the $\langle 110 \rangle$ direction. The spectra for the $\langle 110 \rangle$ and random cases are also shown for comparison. Note that the backscattering yield comes from In only [143]

analysis, the channeling signals must be collected in a backward direction. In this case, ion backscattering cross sections are sometimes too small for practical channeling mapping of reasonable quality. It is of technical interest to investigate the applicability of the binary-encounter electron yield, which reflects the atomic structure of the surface layer up to a depth of 10–100 nm.

The cylindrical-mirror analyzer, described in Sect. 5.1.2, allows effective use of extremely low-current ion beams, down to the picoampere range. This is weaker than the beam current required in most of the work on ion backscattering spectroscopy by a factor of 10^2–10^3. The measurement technique can be applied to channeling microscopy in a backward direction. Some preliminary results are presented in the following.

Figure 10.8 shows an example of the shadowing pattern around the Ge$\langle 111 \rangle$ axis over an angular range of $2° \times 2°$ (tilt angles θ and ϕ, see Sect. 5.3), measured using electrons induced by 1.25 MeV/u O^{4+} collimated with a 25 μm diameter aperture. The measured energy range of $4.1 \leq E/eV_p \leq 8.3$ (the high-energy end of the measurable area shown in Fig 5.4), i.e. $E = 2.0$–4.1 keV, was chosen by setting $V_p = 0.49$ kV and $V_g = 2.0$ kV. In this case, therefore, the electron yield at energies near $E_B = 2.7$ keV was measured. An O^{4+} beam current on the target as low as 7 pA provided a sufficient count rate of ~2000 counts/s under random incidence conditions. The counting time required for each of the 21 × 21 pixels of tilt angle to measure the pattern was ~2 s, which corresponds to a fixed beam dose of 4.5×10^{12} O^{4+}/cm^2.

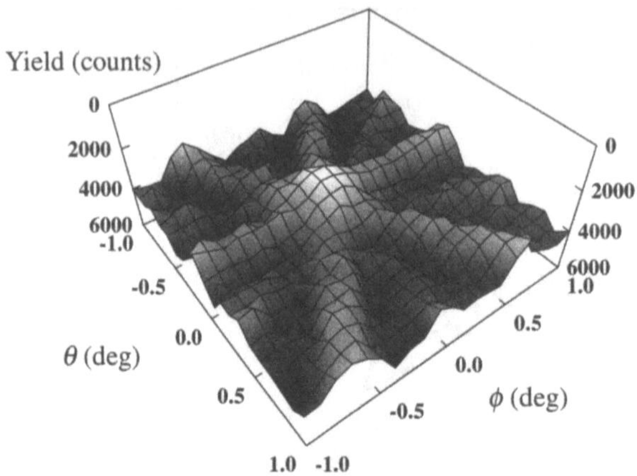

Fig. 10.8. Shadowing pattern around Ge⟨111⟩ over an angular range of 2° × 2°, measured with 2.0–4.1 keV electrons induced by 1.25 MeV/u O^{4+}. Note that the yield axis points downward [70]

If the observation angle is 180° with respect to the ion beam, the random electron yield does not depend on the tilt angle of the crystal, as discussed in Sect. 6.2. For a wide observation angle as in the present case, account must be taken of the tilt-angle dependence of the random electron yield since it could affect the shadowing pattern. In Fig. 10.8, however, no such dependence can be recognized in the angular range of crystal tilt used here, within ~2° from the surface normal.

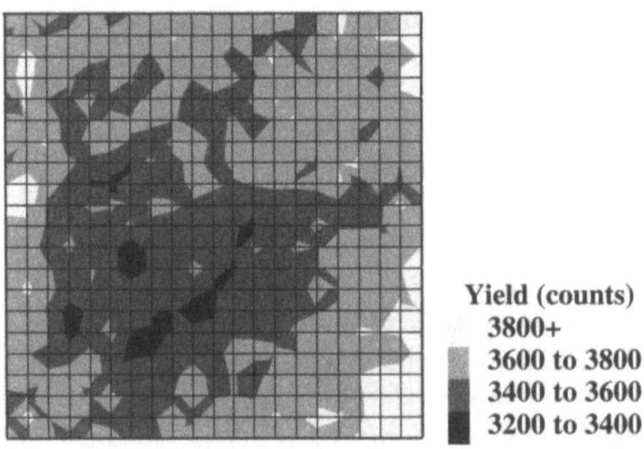

Fig. 10.9. Typical 1 × 1 mm^2 image of the Ni⟨110⟩ electron yield, visualizing the imperfect crystallinity of the Ni crystal [70]

In channeling microscopy, a collimated beam is scanned over a finite area of the crystal surface under channeling incidence conditions. In the work described here, the scanning was performed by means of horizontal and vertical shifts of the crystal with respect to the beam direction. Figure 10.9 shows a $1 \times 1\,\mathrm{mm}^2$ image of an imperfect Ni crystal, measured using the Ni⟨110⟩ electron yield under the same experimental conditions as noted above. The random yield corresponds to ~6000 counts (the normalized ⟨110⟩ yield is ~0.6). The actual spatial resolution is ~60 μm in both the horizontal and the vertical directions; this was estimated from image profiles of sharp wire edges. The resolution is ~2.5 times wider than the aperture diameter, and is probably due to scattering of the ions at the aperture edge. Further measurements of the ⟨110⟩ channeling direction indicate the absence of the macroscopic bending of the crystal within $1 \times 1\,\mathrm{mm}^2$ areas. The imperfect crystallinity of the Ni crystal is visualized well in Fig. 10.9, demonstrating essentially the same kind of information as obtained by ion backscattering analysis. On the other hand, channeling microscopy images of high-crystallinity Si measured under the same experimental conditions showed only uniform contrast within the uncertainty of the electron counts of ~3%.

These preliminary results confirm the feasibiliy of backward channeling microscopy using ion-induced electrons. In the above example, the minimum beam dose required for analysis of the channeling electron yield can be reduced to as low as $\sim 10^{11}$ ions/cm^2. This value is much smaller than the values required in typical ion backscattering spectroscopy measurements (for example, $\sim 10^{14}$ ions/cm^2 for Si targets [43]), by a factor of $\sim 10^3$. Such an experimental condition allows effective use of a tightly collimated picoampere beam. Use of a microbeam should allow technological progress in the spot-area structural analysis of surface layers.

Some other applications of low-beam-current analysis will be briefly noted. These applications could include analyses of radiation-sensitive materials. Electronic excitations caused by an analyzing ion beam induce serious damage in some insulating crystal films, and therefore the usual ion beam analyses are not readily applicable. Furthermore, an electron analyzer of wide acceptance angle might be of practical importance for channeling and related experiments using secondary charged particles generated by accelerators, if the available beam current is at least in the picoampere range.

10.5 Bent Ionic Crystals

The cylindrical-mirror electron analyzer (Sect. 5.1.2) allows low-beam-dose analysis of crystalline materials which suffer serious beam-induced damage under typical beam dose conditions (much higher than 10^{15} ions/cm^2) with MeV ions. Some structural studies of plastically bent ionic crystals are presented to demonstrate the application of low-beam-dose analysis using ion-induced electrons [79].

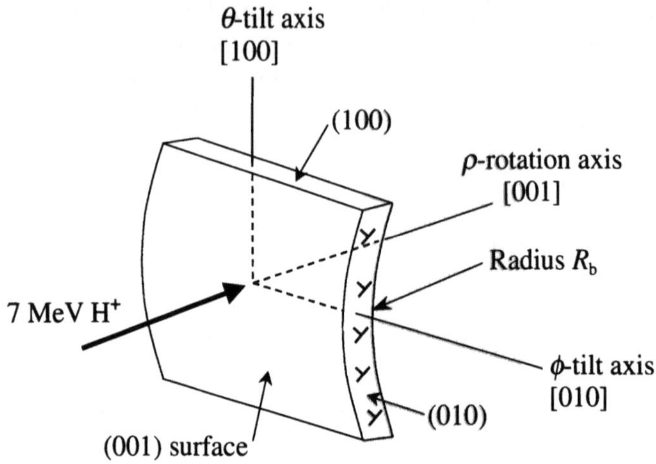

Fig. 10.10. Schematic arrangement of the experiments. A 7 MeV H^+ beam is incident on the (001) surface of a KCl or NaCl sample prepared by cleavage. The surface normal direction is taken to be $[00\bar{1}]$, while the cleaved $\{100\}$ side surfaces are perpendicular to [100] or [010]. The three directions coincide with the ϱ, θ, and ϕ axes for crystal rotation and tilts. Edge dislocations with a Burgers vector $\boldsymbol{b} = (\hat{a}/\sqrt{2})[101]$ are shown by the common symbol. The electron analyzer is placed in front of the (001) surface [79]

10.5.1 Experimental Details

KCl and NaCl samples were prepared by repeated cleavage of crystal blocks to a thickness of 0.2–0.3 mm. The areas of the samples were typically $10 \times 5\,\mathrm{mm}^2$ (KCl) and $5 \times 2\,\mathrm{mm}^2$ (NaCl). As a result of the cleavage process, the samples were bent to form a cylindrically curved surface with a radius in the centimeter range. The bending direction is preferential, so that the cylinder axis is parallel to one of the $\langle 100 \rangle$ edges of the sample, as shown in Fig. 10.10. The samples were fixed on an aluminum sample holder with an electroconductive adhesive and mounted on a three-axis goniometer. The samples were analyzed at room temperature ($\sim$300 K) with a 7 MeV H^+ beam. At this energy the mean penetration length of H^+, roughly 0.5 and 0.4 mm in KCl and NaCl, respectively [155], exceeds the thickness and therefore causes no charging effect in the electron measurements. The beam current on the target was decreased to 300 pA for a beam spot size of $\sim 0.5 \times 0.5\,\mathrm{mm}^2$.

Figure 10.11 shows typical energy spectra of electrons emitted from a bent KCl crystal under [001] and random incidence conditions, measured using a parallel-plate spectrometer of the type shown in Fig. 5.2, at 180° with respect to the beam direction. The [001] yield was dependent on the beam position on the sample surface, reflecting the lattice imperfection due to the bending of the crystal. The normalized [001] yield in the 3–8 keV range ($\sim$0.60 in Fig. 10.11) could be used to investigate the lattice structure of the bent crystals.

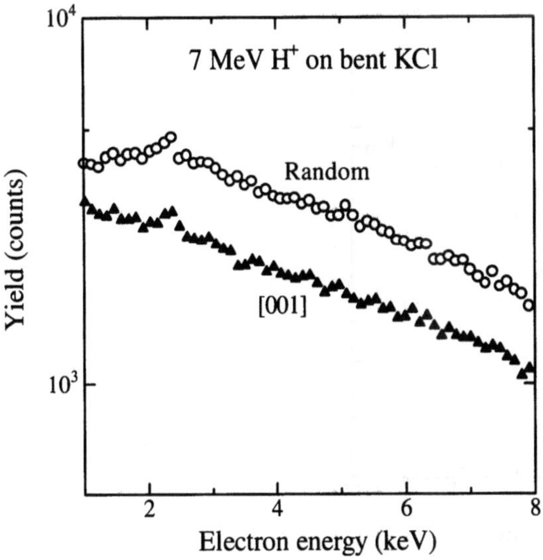

Fig. 10.11. Energy spectra of secondary electrons induced by 7 MeV H^+, measured for a bent KCl crystal at a backward angle of 180° under [001] and random incidence conditions. The relative energy resolution of the parallel-plate spectrometer was $\Delta E/E \simeq 5\%$ [79]

The primary Auger peak due to Cl KLL Auger electrons is seen at 2.4 keV, which agrees with the value predicted from the atomic energy levels of the Cl atom [113]. This confirms the absence of any charging effect, which would otherwise cause a low-energy shift of the Auger peak. Similar spectra were observed for NaCl.

The energy range of the electrons measured in the experiments was 5 to 7 keV, obtained by applying plate and grid voltages of −1 and −5 kV, respectively. This operation condition was chosen so as to obtain a sufficient count rate of the electron yield and to observe the shadowing effect clearly. Note that the above energy range lies below the binary-encounter peak energy $E_B = 15.2$ keV, and therefore the inner-shell as well as the valence electrons contribute to the observed yield.

The shadowing patterns were obtained by measuring the 5–7 keV yield as a function of the angles θ and ϕ shown in Fig. 10.10. For both KCl and NaCl, the beam dose needed to measure a shadowing pattern of 21×21 pixels, which covered an angular range of 2° × 2° in θ and ϕ, was 1.1×10^{14} H^+/cm^2. The measurement time for the pattern was ∼10 min. For ionic crystals, account must be taken of beam-induced damage. Figure 10.12 shows the beam dose dependence of the normalized [001] yield for a bent KCl sample. We see that W increases appreciably after an irradiation dose of $\sim10^{15}$ H^+/cm^2. This is essentially consistent with reported damage studies of KCl and NaCl by Rutherford backscattering spectroscopy using 1 MeV He^+ [156, 157]. Accord-

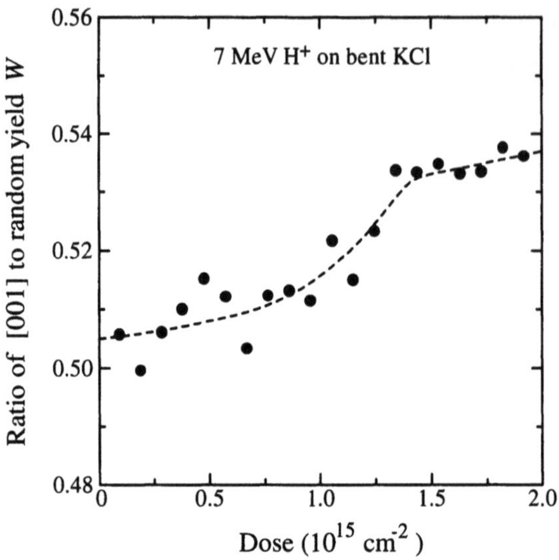

Fig. 10.12. Beam dose dependence of the normalized [001] electron yield at 5–7 keV, measured for a bent KCl crystal. The *dashed curve* is drawn to guide the eye [79]

ingly, the low-beam-dose conditions in the present case ($1.1 \times 10^{14}\,H^+/cm^2$) ensure negligible influence from radiation damage.

The measured electron yield provides structural information from the surface to a depth equal to the effective escape length for the electron yield, D. In the present case, the value of D estimated using (8.2) and the calculated value of t is roughly in the range 500–1000 Å for KCl and NaCl. The value of D is not of essential importance in the following discussion, and therefore we shall not discuss it in more detail.

10.5.2 Abnormal Shadowing Patterns

In the present study, four or five samples of both KCl and NaCl were investigated to deduce the specific features of the bent crystal lattice. Figure 10.13 shows typical shadowing patterns around the [001] axis perpendicular to the bent (001) surfaces of KCl and NaCl. The radii of curvature R_b (Fig. 10.10) were approximately equal to 45 and 5 mm for the KCl and NaCl samples, respectively. A typical observed pattern is shown in Fig. 10.13a, where the (010) plane at $\theta = 0$ can be clearly identified, while the other planes are relatively degraded. Also shown, in Figs. 10.13b, c, are patterns observed near the corners of samples. In Fig. 10.13b, the pattern appears more symmetric than in Fig. 10.13a, but it does not correspond to a defect-free lattice, since even the [001] axial yield at $\theta \simeq -0.2°$, $\phi \simeq -0.3°$ is at approximately the same level as the (010) planar yield at $\theta \simeq 0$ in Fig. 10.13a. In Fig. 10.13c,

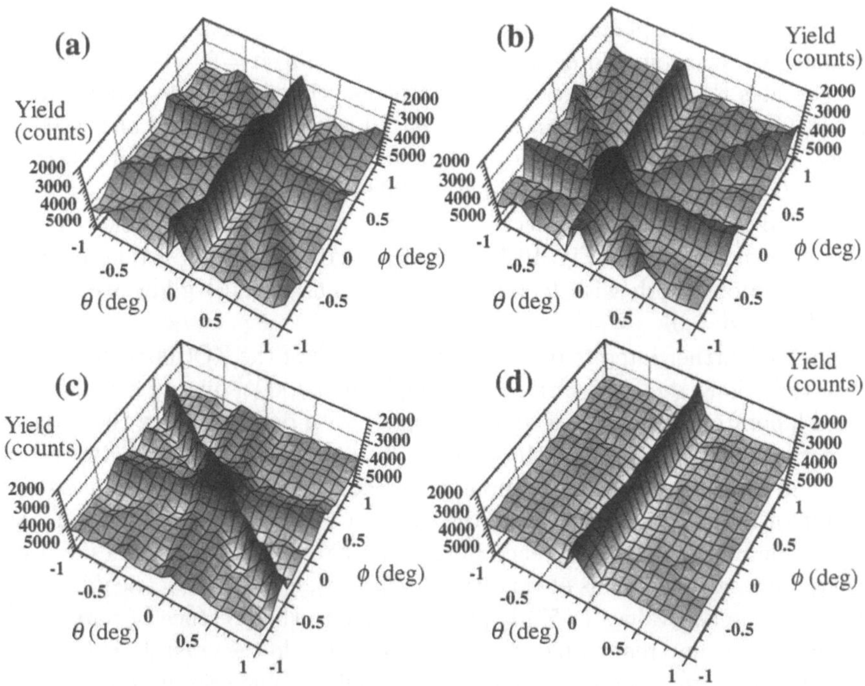

Fig. 10.13. Shadowing patterns around [001] for a bent KCl crystal (**a**, **b**, and **c**) and NaCl crystal (**d**). In (**d**), the surface normal direction (parallel to $[00\bar{1}]$ before bending) corresponds to $\theta \simeq 0°$, $\phi \simeq 0°$. The yield axes point downward [79]

the clear planar image is not that of (010), but of (110), passing through $\theta \simeq 1°$, $\phi \simeq -0.6°$.[2] Furthermore, the (010) pattern at $\theta \simeq 0°$ is slightly split by an angle of $\sim 0.3°$, possibly indicating the existence of a grain boundary. Figure 10.13d shows a pattern for a bent NaCl crystal, where only the (010) planar shadowing at $\theta \simeq -0.2°$ is seen.

The experimental results can be explained in terms of the plastic deformation accompanying edge dislocations. This is typical of simple ionic crystals [158]. In these crystals, the Burgers vector of an edge dislocation is given by $\boldsymbol{b} = (\hat{a}/\sqrt{2})\langle 110\rangle$,[3] and the glide plane is $\{110\}$. When edge dislocations are produced by the cylindrical bending of KCl or NaCl shown in Fig. 10.10, the preferential direction of $\boldsymbol{b}$ is [101], including the equivalent directions of $[\bar{1}01]$, $[10\bar{1}]$, and $[\bar{1}0\bar{1}]$. The Burgers vectors in the other $\langle 110\rangle$ directions gen-

[2] Note that the shadowing pattern in Fig 10.13c was measured after rotating the crystal counterclockwise with respect to the ϱ axis by $\sim 13°$, as can be seen from a comparison of planar directions with other cases.

[3] The lattice constants are $\hat{a} = 6.29$ and 5.64 Å for KCl and NaCl, respectively, at room temperature.

erate atomic displacements with components parallel to [010]. This would degrade the (010) planar pattern, contrary to the observations. Obviously, planar shadowing should be unaffected only when $\boldsymbol{b} \cdot \boldsymbol{n} = 0$, where the vector $\boldsymbol{n}$ indicates the normal direction of the planar channel. In the present case, an (010) shadowing which satisfies $\boldsymbol{b} \cdot \boldsymbol{n} = 0$ is clearly observed at $\theta = 0°$ in Figs. 10.13a, d. The shadowing patterns of somewhat different type seen in Figs. 10.13b, c may be attributed to a different deformation resulting from a relaxed stress field at the corners of the samples during the cleavage. It should be noted that in Fig. 10.13c the clear (110) shadowing is due to a pair of dislocations, both with $\boldsymbol{b} = (\hat{a}/\sqrt{2})\langle 110 \rangle$ plus a $\{110\}$ glide plane, that satisfies $\boldsymbol{b} \cdot \boldsymbol{n} = 0$.

It is of further interest to investigate the shift of the KCl [001] shadowing direction along the bent surface. The angle of the [001] direction relative to fixed laboratory coordinates changes linearly with distance along the vertical direction (~5 mm) in Fig. 10.10, and the [001] angular difference between the top and bottom of the KCl sample is $1.4 \pm 0.1°$, according to the measurements of the [001] angular dip of the electron yield. On the other hand, the corresponding angular difference in the surface normal direction, estimated from the measured radius of curvature (~45 mm), is $6.4 \pm 1.4°$. The difference between the two values (~5°) is probably due to the presence of pure (110) glide without any remaining edge dislocations inside the crystal. This bends the crystal plate without changing the crystallographic orientation.

10.5.3 Plastically Deformed Lattices

The observed anisotropy in the planar shadowing patterns can hardly be accounted for in terms of planar dechanneling by dislocations [43, 55, 57]. The [001] angular shift of $1.4 \pm 0.1°$ for the KCl crystal of 0.2 mm thickness should be accompanied by an expansion of the width of the (001) surface by $\delta x = 4.9 \times 10^{-3}$ mm. The number of (110) extra half-planes needed to relax the expansion can be estimated as $\delta x/(b/\sqrt{2}) = 1.6 \times 10^4$. It follows that the number density of edge dislocations in the (010) surface N_e, shown by the symbols for edge dislocations in Fig. 10.10, is equal to $1.6 \times 10^6\,\mathrm{cm}^{-2}$. The mean diameter of the cylindrical region belonging to a dislocation line is obtained as $2(N_\mathrm{e}\pi)^{-1/2} = 8.9 \times 10^4$ Å. This value is much larger than the effective escape length for the electron yield ($D = 500$–1000 Å) noted earlier, so that the ions experience at most a single encounter with a dislocation. The channeled ions can be scattered by a curved plane of atoms near a dislocation line and become nonchanneled. Such a transition effectively occurs when an ion passes within a finite distance L_d from the dislocation line. The dechanneling width $2L_\mathrm{d}$ can be roughly estimated from the commonly used analytical formula given by Quéré [43, 159]. It depends on the critical angle of channeling, the Burgers vector, and a parameter depending on the type of dislocation. The dechanneling width in the present case is estimated to be ~510 Å. Therefore, even as an overestimate, only a beam fraction of

$510/(8.9 \times 10^4) = 0.6\%$ is expected to be dechanneled and increase the electron yield. As has been pointed out for planar channeling, Quéré's formula tends to underestimate the dechanneling width by a factor of ~0.5 [43, 160]. Even if the estimated dechanneling width includes such an ambiguity, dechanneling by dislocations hardly explains the abnormal shadowing patterns in Fig. 10.13a–c.

The observed anisotropy is mainly due to continuously misoriented (100) planes, caused by the preferential configuration of the edge dislocations. The [001] angular shift of 1.4° along a 5 mm distance over the curved surface implies a possible misorientation of the (100) direction with respect to the beam direction by ~0.14° for the present beam size (~0.5 mm width). This value is greater than the critical angle of ~0.08° for KCl {100}, estimated from the angular width of the pattern shown in Fig. 10.13. As a rough estimate, a beam fraction of $(0.14 - 0.08)/0.14 = 50\%$ might be effectively incident in a random direction, while the other 50% of the ions should be subject to the shadowing effect. Therefore, the normalized (100) yield can be expressed as $0.5(1 + W)$, where W is the corresponding yield for (010). The observed value of $W \simeq 0.51$ for the undegraded (010) yield along $\theta \simeq 0°$ in Fig. 10.13a leads to $0.5(1 + W) \simeq 0.76$. This is in good agreement with the observed value of ~0.8 for the (100) yield along $\phi \simeq 0.3°$ in Fig. 10.13a. For tighter bending, all the planar patterns will completely vanish except for the (010) plane that satisfies $\boldsymbol{b} \cdot \boldsymbol{n} = 0$. The single NaCl(010) pattern shown in Fig. 10.13d certainly corresponds to such a tight bending condition.

The analysis presented here has demonstrated the practical use of ion-induced electron emission for structural analysis with extremely low beam damage. A similar structural analysis might be possible using X-ray diffraction, which appears to be the most competitive technique for characterization of plastically deformed crystals [161]. However, the X-ray method is essentially bulk-sensitive (a sensitive depth of 100 μm or greater), whereas ion beam analysis is sensitive to a submicron or shallower surface layer. Accordingly, the two techniques are complementary. Further technical improvement of the present system should allow structural information to be obtained with a beam dose of as low as $\sim 10^{11}\ H^+/cm^2$ for a $1\ mm^2$ beam spot on the target. This value can be decreased for equal-velocity ions of atomic number Z_1 by a factor of Z_1^{-2}, since the number of electrons measured increases according to the Z_1^2 scaling of the binary-encounter electron yield.

11. Related Topics

This chapter presents some other topics related to the ion-induced electron emission from crystalline solids. Most of them are concerned with low-energy electron emission from crystal surfaces. The production and escape processes of low-energy electrons have been discussed extensively in a number of review articles cited in Chap. 1, and are beyond the scope of this monograph. The following discussion is therefore focused only on the kind of knowledge obtainable from experiments using crystal targets.

11.1 Channeling Angular Dips

The total electron yield emitted from an ion-bombarded surface is dominantly composed of low-energy electrons of kinetic energies lower than ~50 eV [1,2]. These electrons are effectively emitted from a surface layer whose thickness is equal to an effective escape length D_{L}, which is at most of the order of 10 Å. The key factor for the appearance of channeling dips in the tilt-angle dependence of the total electron yield is obviously the relative balance between the effective surface layer thickness t (for shadowing of the outer shell of the atom) and D_{L}. Channeling dips will not be observed in the case $t \geq D_{\mathrm{L}}$, since most of the observed electrons originate from the unshadowed surface layer, and therefore a shadowing effect is not expected.

The validity of this criterion was confirmed experimentally by Hasegawa et al. [162]. They observed clear channeling dips for 1–3 keV/u Na^+ incident on an SnTe crystal, while for 250 keV/u He^+ the dips were seriously degraded or even disappeared. For Na^+ incident ions, mainly the surface atoms are exposed to the beam, so that effectively $t \sim 0$. In this case, the condition for observation of dips, $t < D_{\mathrm{L}}$, is satisfied. It should be noted that earlier observations of channeling dips were carried out under similar experimental conditions [8, 10, 11]. For He^+ incident ions, however, the increased ion velocity leads to an enhanced value of $t \gtrsim 50$ Å, which exceeds D_{L} for typical conditions of axial or planar incidence on SnTe, according to the calculations of Hasegawa et al. This is a typical condition for the absence of channeling dips, and is consistent with observation.

11.2 Specularly Reflected Ions

Fast ions incident at a grazing angle on a crystal surface can be specularly reflected if the incidence angle relative to the surface is smaller than $\sim\psi_p$, as given by (4.40), and the incident direction is not parallel to any low-index axial direction on the surface. Specular reflection and channeling in a surface layer were comprehensively studied at the University of Sussex around 1970 [163]. Later, specular reflection has provided an attractive experimental technique for observing ion–surface interactions, for example image charge effects of energetic ions moving along solid surfaces [164, 165].

By changing the grazing incidence angle on the surface, the closest distance of approach to the top surface layer, d_{min}, can be varied. Hence the impact parameter dependence of ion-induced electron emission from the surface atoms can be studied. Applying this technique, Pfandzelter and Landskron studied Cu $M_{2,3}VV$ Auger electron emission from Cu(111) bombarded by 175 keV protons [166]. Auger emission exclusively from the Cu(111) top surface layer was observed for $d_{min} \simeq 0.9$ Å, corresponding to a grazing incidence angle of 0.5°, while the Auger intensity was reduced for $d_{min} \simeq 1.9$ Å by a factor of 0.2 relative to that for $d_{min} \simeq 0.9$ Å. The reduced Auger intensity is attributed to the reduced ionization probability for the Cu M shell at a large impact parameter. The ultimate surface sensitivity of one monolayer obtained in this experiments cannot be achieved using an electron beam.

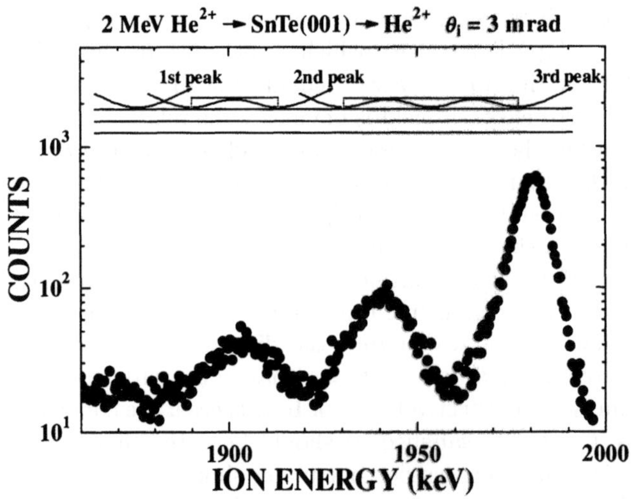

Fig. 11.1. Energy spectrum of 2 MeV He^{2+} specularly reflected from SnTe(001) [169]

Studies of low-energy electron emission induced by specularly reflected ions have been reported in several papers [77, 78, 167–170]. In the experiments

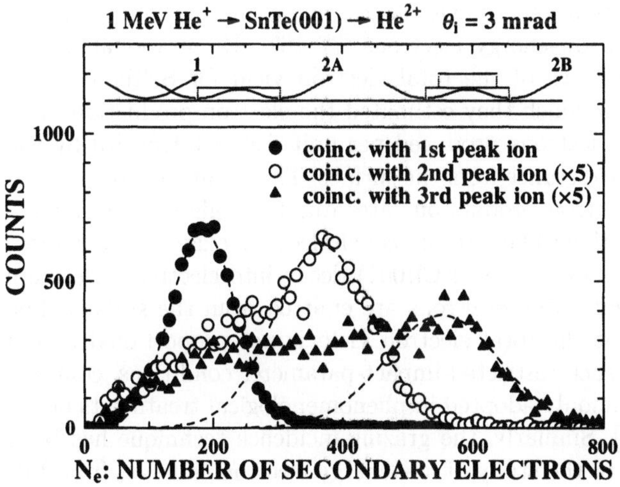

Fig. 11.2. Number distribution of secondary electrons emitted from SnTe(001) for grazing incidence of 1 MeV He^+, measured in coincidence with specularly reflected He^{2+} suffering energy losses corresponding to $n_c = 1, 3, 5$ [169]

of Kimura et al., specularly reflected ions and subsurface channeled ions were distinguished in the energy spectrum of ions reflected from the (001) surface of an epitaxially grown SnTe layer. In Fig. 11.1, the three pronounced peaks in the spectrum, measured for 3 mrad grazing incidence of 2 MeV He^{2+}, result from energy losses of $\sim 20 \times n_c$ keV ($n_c = 1, 3, 5$). The presence of loss peaks only for odd values of n_c certainly indicates the existence of a step structure on the surface [165]. In fact, n_c is the number of glancing collisions with the (001) sheet of atoms, as shown schematically in the inset of Fig. 11.1. Kimura et al. also studied the n_c dependence of the number of emitted low-energy electrons from the measurements in coincidence with ions at the loss peaks for $n_c = 1, 3, 5$. The experimental results are shown in Fig. 11.2. In this case, the pulse-height distribution of the output signals from the microchannel plate corresponds to the number distribution of the emitted secondary electrons. The peaks of the three fitted Gaussians shown by the dashed curves give the most probable numbers of the low-energy electrons produced, $N_e \simeq 183$, 369, and 553. It is evident that the most probable number is proportional not to n_c, but to $(n_c + 1)/2$, i.e. the effective number of glancing collisions with the (001) sheet of atoms that are exposed to the vacuum; see the inset in Fig. 11.1. At a pyramidal hillock, shown in the inset (case 2B) in Fig. 11.2, an appreciable number of outgoing electrons should fail to escape through the overlaid (001) sheets of atoms, giving rise to the low-number tails of the peaks for $(n_c + 1)/2 = 2$ and 3 (corresponding to the second and third peaks, as indicated in Fig. 11.1).

The specular-reflection experiments also provide information about the producion processes of low-energy electrons [77, 167]. Kimura et al. determined the d_{min} dependence of the total electron yield for SnTe(001) and KCl(001) surfaces, from which they estimated the electron production rates (number of secondary electrons produced per unit distance traveled by the ion) as a function of the distance from the surface. The results were compared with the calculated electron production rates due to single-electron ionization or plasmon excitation. The authors concluded, for example, that most of the surface plasmons excited on KCl(001) decay into electron–hole pairs and that half of the generated electrons are emitted from the surface. The authors also found that the total electron yield is not proportional to the stopping power under such restricted-impact-parameter conditions, contrary to the assumption commonly adopted in phenomenological treatments of the total electron yield [3]. Similarly, the grazing-incidence technique has been applied by Eder et al. to the study of low-energy electron emission from LiF bombarded by 1–20 keV H^0 (hydrogen) [78].

When ions in the eV/u energy range are incident on a solid surface, the dominant mechanism for electron emission is so-called *potential emission* [3]. In this case, electron transitions between the electronic energy levels of the projectile and the solid provide the kinetic energy of the emitted electrons. As the ion energy increases, the main electron emission mechanism shifts to kinetic processes (*kinetic emission*) such as binary-encounter electron emission. For ions in the keV/u energy range, the two emission mechanisms are competitive. Lemell et al. studied low-energy electron emission from Au(111) bombarded by 0.45 keV/u Ar^{8+} under grazing incidence (~5°) conditions [168]. In these experiments, the number distributions of emitted electrons were measured in coincidence with scattered ions at a fixed observation angle of interest. The number distribution was clearly dependent on the observation angle. At the specular reflection angle, the number distribution became identical to that for normal incidence of 2.5 eV/u Ar^{8+}, whose velocity is approximately equal to the normal velocity component of the grazingly incident 0.45 keV/u Ar^{8+} used. Thus these authors found that kinetic emission by specularly reflected ions is considerably suppressed relative to potential emission.

11.3 Low-Energy Shadowing

It is of fundamental interest to follow the variation of all processes of ion-induced electron emission from shadowed crystals as the ion energy decreased to 1 keV/u or lower. The individual surface atoms cast a wide shadow for low-energy ions. In fact, the critical angle corresponding to ψ_A given by (4.25) for the high-energy case is of the order of 10° in low-energy ion scattering spectroscopy. Accordingly, a reduced total electron yield can be observed when ions are incident on a surface at an angle less than the critical angle relative to an axial direction on the surface. The observed incidence-angle dependence can be directly converted to the impact-parameter dependence of the total electron yield. Rabalais et al. thus measured the incidence-angle dependence of the total electron yield emitted from Ni(110) bombarded by 4 keV Ar^+ by a time-of-flight method. They found that the emission of secondary electrons requires Ar–Ni collisions with impact parameters less than a threshold value of 0.3 Å [171]. This value is approximately equal to the Ar–Ni spacing needed for significant overlap of the Ar and Ni L shells, which allows electron emission to be caused by electron promotion in molecular orbitals formed during the collision.

12. Concluding Remarks

This monograph has the intention of revealing not only fundamental, but also applied aspects of the spectroscopy of ion-induced energetic electrons that are produced by the binary-encounter processes in solids bombarded by fast ions. The binary-encounter electron yield considered here results from multiple scattering of energetic ions and electrons, which would require numerical calculations in an *orthodox* approach. It has been demonstrated, however, that the electron yield under channeling as well as random incidence conditions can be explained by the model based on a useful concept, i.e. the effective surface layer for production of the electron yield, which was deduced from the experimental data covering wide ranges of the ion energy and ion species.

The analysis model of the binary-encounter electron yield discussed in this book certainly enables insight into the seemingly complicated scattering and outgoing processes of energetic electrons produced in crystalline or non-crystalline solids. Also, the experimental technique developed is an example of a successful combination of the well-established experimental techniques, i.e. ion channeling and electron spectroscopy, which have so far rarely been used in this way.

References

1. J. Schou: Scann. Microsc. **2**, 44 (1988)
2. M. Rosler, W. Brauer, J. Devooght, J.-C. Dehaes, A. Debus, M. Cailler, J.-P. Ganachaud: *Particle Induced Electron Emission I* (Springer, Berlin, Heidelberg, 1991)
3. D. Hasselkamp, H. Rotherd, K.-O. Groeneveld, J. Kemmler, P. Varga, H. Winter: *Particle Induced Electron Emission II* (Springer, Berlin, Heidelberg, 1991)
4. R. A. Baragiola: "Electron emission from slow ion–solid interactions". In: *Low Energy Ion-Surface Interactions*, ed. by J. W. Rabalais (Wiley, New York, 1994) pp. 187–262
5. K. Koyama: "Secondary electron emission". In: *Ion Beam Technology*, ed. by F. Fujimoto, K. Komaki (Uchida-roukakuho, Tokyo, 1995, in Japanese) Sect. 6.4
6. M. E. Rudd, Y.-K. Kim, D. H. Madison, T. J. Gay: Rev. Mod. Phys. **64**, 441 (1992)
7. N. Stolterfoht, R. D. DuBois, R. D. Rivarola: *Electron Emission in Heavy Ion–Atom Collisions* (Springer, Berlin, Heidelberg, 1997)
8. E. S. Mashkova, V. A. Morchanov, D. D. Odintsov: Dokl. Akad. Nauk. SSSR **151**, 1074 (1963) [Sov. Phys. Dokl. **8**, 806 (1964)]
9. G. D. Magnuson, C. E. Carlston: Phys. Rev. **129**, 2409 (1963)
10. N. Colombie, B. Fagot, C. Fert: Radiat. Eff. **2**, 31 (1969)
11. B. A. Brusilovsky, V. A. Molchanov: Radiat. Eff. **23**, 135 (1974)
12. N. Benazeth: Nucl. Instrum. Methods **194**, 405 (1982)
13. J. Lindhard: Mat. Fys. Medd. Dan. Vid. Selsk. **34**, No. 14, 1 (1965)
14. D. V. Morgan, ed.: *Channeling* (Wiley, London, 1973)
15. D. S. Gemmell: Rev. Mod. Phys. **46**, 129 (1974)
16. V. E. Yurasova, V. M. Buchanov, M. Golo: Phys. Stat. Sol. **17K**, 187 (1966)
17. M. Negre, J. Mischler, N. Benazeth, D. Spanjaard, Surf. Sci. **78**, 174 (1978)
18. J. Mischler, N. Benazeth: Surf. Sci. **40**, 311 (1973)
19. N. Benazeth, J. Agusti, C. Benazeth, J. Mischler, L. Viel: Nucl. Instrum. Methods **132**, 477 (1976)
20. R. J. MacDonald, L. C. Feldman, P. J. Silverman, J. A. Davies, K. Griffiths, T. E. Jackman, P. R. Norton, W. N. Unertl: Nucl. Instrum. Methods **218**, 765 (1983)
21. M. Schuster, C. Varelas: Surf. Sci. **134**, 195 (1983)
22. H. Kudo, D. Schneider, E. P. Kanter, P. W. Arcuni, E. A. Johnson: Phys. Rev. B **30**, 4899 (1984)
23. J. J. Thomson: Phil. Mag. **23**, 449 (1912)

24. C. Lehmann: *Interaction of Radiation with Solids and Elementary Defect Production* (North-Holland, Amsterdam, 1977)
25. N. Stolterfoht: "Excitation in energetic ion–atom collisions accompanied by electron emission". In: *Structure and Collisions of Ions and Atoms*, ed. by I. A. Sellin (Springer, Berlin, Heidelberg, 1978) pp. 155–199
26. E. Gerjuoy: Phys. Rev. **148**, 54 (1966)
27. L. H. Thomas: Proc. Cambridge Philos. Soc. **23**, 713 (1927)
28. M. Gryziński: Phys. Rev. **115**, 374 (1959)
29. M. Gryziński: Phys. Rev. **138**, A305, A322, A336 (1965)
30. L. Vriens: Proc. Phys. Soc. (London) **90**, 935 (1967)
31. J. D. Garcia: Phys. Rev. **177**, 223 (1969)
32. V. Fock: Z. Physik **98**, 145 (1935)
33. M. E. Rudd, D. Gregoire, J. B. Crooks: Phys. Rev. A **3**, 1635 (1971)
34. M. Inokuti: Rev. Mod. Phys. **43**, 297 (1991)
35. K. Takayanagi: *Collisions of Electrons, Atoms and Molecules*, (Baifukan, Tokyo, 1972, in Japanese) Chaps. 4, 5
36. N. F. Mott, H. S. Massey: *The Theory of Atomic Collisions*, 3rd ed. (Oxford University Press, Oxford, 1965) Chap. 3
37. L. I. Schiff: *Quantum Mechanics*, 3rd ed. (McGraw-Hill, New York, 1968) Chap. 5
38. C. O. Reinhold, D. R. Schultz, R. E. Olson: J. Phys. B **23**, L591 (1990)
39. P. Hvelplund, H. Tawara, K. Komaki, Y. Yamazaki, K. Kuroki, H. Watanabe, K. Kawatsura, M. Sataka, M. Imai, Y. Kanai, T. Kambara, Y. Awaya: J. Phys. Soc. Jpn. **60**, 3675 (1991)
40. F. Biggs, L. B. Mendelsohn, J. B. Mann: At. Data Nucl. Data Tables **16**, 201 (1975)
41. M. E. Rudd, J. H. Macek: Case Studies At. Phys. **3**, 47 (1972)
42. N. Bohr: Mat. Fys. Medd. Dan. Vid. Selsk. **18**, No. 8, 1 (1948)
43. L. C. Feldman, J. W. Mayer, S. T. Picraux: *Materials Analysis by Ion Channeling* (Academic Press, New York, 1982)
44. H. Kudo, K. Shima, T. Ishihara: Phys. Rev. B **47**, 27 (1993)
45. M. Aono: Nucl. Instrum. Methods B **2**, 374 (1984)
46. R. S. Williams: "Quantitative intensity analysis of low-energy scattering and recoiling from crystal surfaces": In: *Low Energy Ion–Surface Interactions*, ed. by J. W. Rabalais (Wiley, New York, 1994) pp. 1–54
47. P. G. Bertrand, J. W. Rabalasi: "Ion scattering and recoiling for elemental analysis and structure determination". In: *Low Energy Ion–Surface Interactions*, ed. by J. W. Rabalais (Wiley, New York, 1994) Chap. 2, pp. 55–116
48. R. Smith, ed.: *Atomic and Ion Collisions in Solids and at Surfaces* (Cambridge University Press, Cambridge, 1997)
49. O. B. Firsov: Zh. Eksp. Teor. Fiz. **33**, 696 (1957) [Sov. Phys. JETP **6**, 534 (1958)]
50. O. S. Oen: Surf. Sci. **131**, L407 (1983)
51. D. J. O'Conner, R. J. MacDonald: "Low energy ion scattering from surfaces". In: *Ion Beams for Material Analysis*, ed. by J. R. Bird, J. S. Williams (Academic Press, Sydney, 1989) Chap. 8
52. A. H. Sørensen: Nucl. Instrum. Methods B **119**, 1 (1996)
53. V. M. Biryukov, Y. A. Chesnokov, V. I. Kotov: *Crystal Channeling and Its Application at High-Energy Accelerators* (Springer, Berlin, Heidelberg, 1997)
54. J. H. Barrett: Phys. Rev. B **3**, 1527 (1971)

55. K. Kimura, T. Oshiyama, M. Mannami: Jpn. J. Appl. Phys. **21**, 1222 (1982)
56. K. Kimura, S. Sawada, M. Mannami: Jpn. J. Appl. Phys. **21**, 1228 (1982)
57. M. L. Swanson: "Channeling". In: *Handbook of Modern Ion Beam Materials Analysis*, ed. by J. R. Tesmer, M. Nastasi (Materials Research Society, Pittsburgh, 1995) Chap. 10
58. B. A. Davidson, L. C. Feldman, J. Bevk, J. P. Mannaerts: Appl. Phys. Lett. **50**, 135 (1987)
59. J. M. Hansteen, O. M. Johnsen, L. Kockbach: At. Data Nucl. Data Tables **15**, 305 (1975)
60. J. H. McGuire: Phys. Rev. A **9**, 286 (1974)
61. J. H. Barrett, D. P. Jackson: Nucl. Instrum. Methods **170**, 115 (1980)
62. D. P. Jackson, T. E. Jackman, J. A. Davies, W. N. Unertl, P. R. Norton: Surf. Sci. **126**, 226 (1983)
63. J. S. Williams, R. G. Elliman: "Channeling". In: *Ion Beams for Material Analysis*, ed. by J. R. Bird, J. S. Williams (Academic Press, Sydney, 1989) Chap. 6
64. D. Roy, J. D. Carette: "Electon spectroscopy for surface analysis". In: *Topics in Current Physics 4*, ed. by H. Ibach (Springer, Berlin, Heidelberg, 1977) Chap. 2.
65. D. L. Matthews: "Ion-induced Auger electron spectroscopy". In: *Methods of Experimental Physics*, Vol. 17, ed. by P. Richard, (Academic Press, New York, 1980) Chap. 9
66. H. Kudo, K. Murakami, K. Takita, K. Masuda, S. Seki, K. Shima, H. Itoh, T. Ipposhi: Jpn. J. Appl. Phys. **24**, 1440 (1985)
67. H. Kudo, K. Shima, S. Seki, K. Takita, K. Masuda, K. Murakami, T. Ipposhi: Phys. Rev. B **38**, 44 (1988)
68. A. Sakamoto, H. Kudo, T. Ishihara, S. Seki, K. Sumitomo: Nucl. Instrum. Methods B **140**, 47 (1998)
69. H. Kudo, A. Tanabe, T. Ishihara, S. Seki, Y. Aoki, S. Yamamoto, P. Goppelt-Langer, H. Takeshita, H. Naramoto: Nucl. Instrum. Methods B **115**, 125 (1996)
70. H. Kudo, K. Takada, K. Narumi, S. Yamamoto, H. Naramoto, S. Seki: Nucl. Instrum. Methods B **142**, 402 (1998)
71. F. Bordoni: Nucl. Instrum. Methods **97**, 405 (1971)
72. L. H. Toburen: Phys. Rev. A **3**, 216 (1971)
73. T. Kuroiwa: B. Sc. Thesis, University of Tsukuba, 1995
74. M. Galanti, R. Gott, J. R. Renaud: Rev. Sci. Instrum. **42**, 818 (1971)
75. Hamamatsu Photonics: MCP Assembly, Technical Information (No. TMCP9001E02), May 1994, Shizuoka, Japan
76. H. Kudo, K. Shima, K. Takita, K. Masuda, K. Murakami, H. Itoh, T. Ipposhi, S. Seki: Jpn. J. Appl. Phys. **25**, 1751 (1986)
77. K. Kimura, G. Andou, K. Nakajima: Phys. Rev. Lett. **24**, 5438 (1998)
78. H. Eder, A. Mertens, K. Maass, H. Winter, HP. Winter, F. Aumayr: Phys. Rev. B **62**, 052901 (2000)
79. H. Kudo, T. Kumaki, K. Haruyama, Y. Tsukamoto, S. Seki, H. Naramoto: Nucl. Instrum. Methods B **174**, 512 (2001)
80. T. Matsukawa, R. Shimizu, H. Hashimoto: J. Phys. D: Appl. Phys. **7**, 695 (1974)
81. R. Shimizu: Jpn. J. Appl. Phys. **22**, 1631 (1983)
82. S. Valkealahti, R. M. Nieminen: Appl. Phys. A **32**, 95 (1983)
83. S. Valkealahti, R. M. Nieminen: Appl. Phys. A **35**, 51 (1984)

84. H. Kudo, S. Seki, K. Sumitomo, K. Narumi, S. Yamamoto, H. Naramoto: Nucl. Instrum. Methods B **164–165**, 897 (2000)
85. H. Kudo, N. Nakamura, K. Shibuya, K. Narumi, S. Yamamoto, H. Naramoto, K. Sumitomo, S. Seki: Nucl. Instrum. Methods B **168**, 181 (2000)
86. H. Kudo, K. Shima, K. Masuda, S. Seki: Phys. Rev. B **43**, 12729 (1991)
87. H. D. Betz: Rev. Mod. Phys. **44**, 465 (1972)
88. I. A. Sellin: Nucl. Instrum. Methods B **10/11**, 156 (1990)
89. W. F. Egelhoff, Jr.: "X-ray photoelectron and Auger electron forward scattering: a structural diagnostic for epitaxial thin films". In: *Ultrathin Magnetic Structures I*, ed. by J. A. C. Bland, B. Heinrich (Springer, Berlin, Heidelberg, 1994) pp. 220–303
90. H. C. Poon, S. Y. Tong: Phys. Rev. B **30**, 6211 (1984)
91. T. Greber, J. Osterwalder, D. Naumovic, A. Stuck, S. Hufner, L. Schlapbach: Phys. Rev. Lett. **69**, 1947 (1992)
92. H. Hofsäss, G. Linder: Phys. Rep. **201**, 121 (1991)
93. H. Kudo, K. Shibuya, N. Nakamura, T. Azuma, S. Seki: Nucl. Instrum. Methods B **159**, 241 (1999)
94. K. Shibuya: M. Sc. Thesis, University of Tsukuba, 1999
95. H. Kudo, S. Yamamoto, K. Narumi, Y. Aoki, H. Naramoto: Nucl. Instrum. Methods B **132**, 41 (1997)
96. W. F. Egelhoff, Jr.: Phys. Rev. B **30**, 1052 (1984)
97. C. J. Tung, J. C. Ashley, R. H. Ritchie: Surf. Sci. **81**, 427 (1979)
98. H. Li, B. P. Tonner: Phys. Rev. B **37**, 3959 (1988)
99. P. F. A. Alkemade, L. Flinn, W. N. Lennard, I. V. Mitchell: Phys. Rev. A **53**, 886 (1996)
100. R. W. Fink, R. C. Jopson, H. Mark, C. D. Swift: Rev. Mod. Phys. **38** 513 (1966)
101. S. Tougaard, P. Sigmund: Phys. Rev. B **25**, 4452 (1982)
102. M. E. Riley, C. J. MacCallum, F. Biggs: At. Data Nucl. Data Tables **15**, 443 (1975)
103. J. Szajman, R. C. G. Leckey: J. Electron Spectrosc. Relat. Phenom. **23**, 83 (1981)
104. A. T. Tofterup: Phys. Rev. B **32**, 2808 (1985)
105. F. W. Garber, M. Y. Nakai, J. A. Harter, R. D. Birkhoff: J. Appl. Phys. **42**, 1149 (1971)
106. H.-J. Fitting: J. Phys. D **8**, 1480 (1975)
107. L. H. Toburen, W. E. Wilson, H. G. Paretzke: Phys. Rev. **25**, 713 (1982)
108. H. Paul, O. Bolik: At. Data Nucl. Data Tables **54**, 75 (1993)
109. J. F. Ziegler: *Stopping Cross Sections for Energetic Ions in All Elements* (Pergamon, New York, 1980)
110. J. F. Ziegler, J. M. Manoyan: Nucl. Instrum. Methods B **35**, 215 (1988)
111. E. Rauhala: "Energy loss". In: *Handbook of Modern Ion Beam Materials Analysis*, ed. by J. R. Tesmer, M. Nastasi (Materials Research Society, Pittsburgh, 1995) Chap. 2
112. M. H. Brodsky, D. Kaplan, J. F. Ziegler: Appl. Phys. Lett. **21**, 305 (1972)
113. J. A. Bearden, A. F. Burr: Rev. Mod. Phys. **39**, 125 (1967)
114. C. Froese Fischer: At. Data **4**, 301 (1972)
115. E. Lugujjo, J. W. Mayer: Phys. Rev. B **7**, 1782 (1973)
116. S. Datz, F. W. Martin, C. D. Moak, B. R. Appleton, L. B. Bridwell: Radiat. Eff. **12**, 163 (1972)

117. K. Shima, T. Mikumo, H. Tawara: At. Data Nucl. Data Tables **34**, 357 (1986)
118. H. Kudo, T. Fukusho, T. Ishihara, H. Takeshita, Y. Aoki, S. Yamamoto, H. Naramoto: Phys. Rev. A **50**, 4049 (1994)
119. A. W. Hunt, D. B. Cassidy, F. A. Selim, R. Haakennaasen, T. E. Cowan, R. H. Howell, K. G. Lynn, J. A. Golovchenko: Nature **402**, 157 (1999)
120. R. A. Carrigan, J. A. Ellison, eds.: *Relativistic Channeling* (Plenum, New York, 1987)
121. A. Baurichter, C. Biino, M. Clément, N. Doble, K. Elsener, G. Fidecaro, A. Freund, L. Gatignon, P. Grafström, M. Gyr, M. Hage-Ali, W. Herr, P. Keppler, K. Kirsebom, J. Klem, J. Major, R. Medenwaldt, U. Mikkelsen, S. P. Møller, P. Siffert, E. Uggerhøj, Z. Z. Vilakazi, E. Weisse: Nucl. Instrum. Methods B **164-165**, 27 (2000)
122. J. Eichler, W. Meyerhof: *Relativistic Atomic Collisions* (Academic Press, New York, 1995) Chap. 3
123. V. Lottner, R. Sizmann, F. Fujimoto: Nucl. Instrum. Methods B **33**, 73 (1988)
124. H. Kudo, K. Shima, S. Seki, T. Ishihara: Phys. Rev. B **43**, 12736 (1991)
125. K. Shima, T. Ishihara, T. Momoi, T. Mikumo: Phys. Rev. A **29**, 1763 (1984)
126. H. D. Betz: "Charge equilibration of high-velocity ions in matter". In: *Methods of Experimental Physics 17*, ed. by J. N. Mundy, S. J. Rothman, M. J. Fluss, L. C. Smedskjaer (Academic Press, New York, 1980) Chap. 3
127. K. Shima, N. Kuno, M. Yamanouchi: Phys. Rev. A **40**, 3557 (1989)
128. K. Shima, N. Kuno, M. Yamanouchi, H. Tawara: At. Data Nucl. Data Tables **51**, 173 (1992)
129. R. W. Martin: Phys. Rev. Lett. **22**, 329 (1969)
130. H. O. Lutz, S. Datz, C. D. Moak, T. S. Noggle: Phys. Lett. A **33**, 309 (1970)
131. C. Kelbch, S. Hagmann, S. Kelbch, R. Mann, R. E. Olson, S. Schmidt, H. Schmidt-Böcking: Phys. Lett. A **139**, 304 (1989)
132. P. Richard, D. H. Lee, T. J. M. Zouros, J. M. Sanders, J. L. Sinpaugh: J. Phys. B **23**, L213 (1990)
133. K. Taulbjerg: J. Phys. B **23**, L761 (1990)
134. C. O. Reinhold, D. R. Schultz, R. E. Olson, C. Kelbch, R. Koch, H. Schmidt-Böcking: Phys. Rev. Lett. **66**, 1842 (1991)
135. D. R. Schultz, R. E. Olson: J. Phys. B **24**, 3409 (1991)
136. M. Sataka, M. Imai, Y. Yamazaki, K. Komaki, K. Kawatsura, Y. Kanai, H. Tawara: Nucl. Instrum. Methods B **79**, 81 (1993)
137. T. Inoue, H. Kudo, T. Fukusho, T. Ishihara, T. Ohsuna: Jpn. J. Appl. Phys. **33**, L139 (1994)
138. H. Kudo, T. Fukusho, A. Tanabe, T. Ishihara, T. Inoue, M. Satoh, Y. Yamamoto, S. Seki: Jpn. J. Appl. Phys. **34**, 615 (1995)
139. H. Kudo, A. Sakamoto, S. Yamamoto, Y. Aoki, H. Naramoto, T. Inoue, M. Satoh, Y. Yamamoto, K. Umezawa, S. Seki: Jpn. J. Appl. Phys. **35**, L1538 (1996)
140. S. Yamamoto, H. Naramoto, K. Narumi, B. Tsuchiya, Y. Aoki, H. Kudo: Nucl. Instrum. Methods B **134**, 400 (1998)
141. T. Inoue, Y. Yamamoto, S. Koyama, Y. Ueda: Appl. Phys. Lett. **56**, 1332 (1990)
142. T. Inoue, M. Osonoe, H. Tohda, M. Hiramatsu, Y. Yamamoto, A. Yamanaka, T. Nakayama: Appl. Phys. Lett. **59**, 3604 (1991)

143. H. Kudo, K. Shima, T. Ishihara, S. Seki: Jpn. J. Appl. Phys. **29**, L2137 (1990)
144. A. Bontemps, J. Fontenille, A. Guivarc'h: Phys. Lett. A **55**, 373 (1976)
145. A. Bontemps, J. Fontenille: Phys. Rev. B **18**, 6302 (1978)
146. J. U. Andersen, N. G. Chechenin, Z. Z. Hua: Appl. Phys. Lett. **39**, 758 (1981)
147. J. U. Andersen, N. G. Chechenin, Z. H. Zhang: Nucl. Instrum. Methods **194**, 129 (1982)
148. A. D. F. Kahn, J. A. Eades, L. T. Romano, S. I. Shah, J. E. Greene: Phys. Rev. Lett. **58**, 682 (1987)
149. P. J. M. Smulders, D. O. Boerma, M. Shaanan: Nucl. Instrum. Methods B **45**, 450 (1990)
150. L. S. Wielunski, M. S. Kwietniak, G. N. Pain, C. J. Rossouw: Nucl. Instrum. Methods B **45**, 459 (1990)
151. T. Haga, H. Suzuki, M. H. Rashid, Y. Abe: Appl. Phys. Lett. **52**, 200 (1988)
152. S. Adachi: J. Electrochem. Soc. **129**, 609 (1982)
153. J. C. Delsarte, J. C. Jousset, J. Mory, Y. Quéré: In: *Proc. Conf. on Atomic Collision Phenomena in Solids, Brighton, England*, ed. by D. W. Palmer, M. W. Thompson, P. D. Townsend (North-Holland, Amsterdam, 1970) pp. 501
154. P. J. C. King, M. B. H. Breese, P. R. Wilshaw, G. W. Grime: Phys. Rev. B **51**, 2732 (1995)
155. U. Littmark, J. F. Ziegler: *Handbook of Range Distributions for Energetic Ions in All Elements* (Pergamon, New York, 1980)
156. M. J. Hollis: Phys. Rev. B **8**, 931 (1973)
157. M. J. Hollis, C. S. Newton, P. B. Price: Phys. Lett. A **44**, 243 (1973)
158. M. T. Sprackling: *The Plastic Deformation of Simple Ionic Crystals* (Academic Press, London, 1976) Chaps. 2, 4, 6, 10
159. Y. Quéré: Phys. Stat. Sol. **30**, 713 (1968)
160. M. Mannami, K. Kumura, A. Kyoshima, M. Matsusita, N. Natsuaki: J. Phys. Soc. Jpn. **49**, 2319 (1980)
161. H. Naramoto, K. Kamada, Jpn. J. Appl. Phys. **17**, 1915 (1978)
162. M. Hasegawa, T. Fukuchi, Y. Susuki, S. Fukui, K. Kimura, M. Mannami: Jpn. J. Appl. Phys. **30**, 2074 (1991)
163. B. W. Farmery, A. D. Marwick, M. W. Thompson: In: *Proc. Conf. on Atomic Collision Phenomena in Solids, Brighton, England*, ed. by D. W. Palmer, M. W. Thompson, P. D. Townsend (North-Holland, Amsterdam, 1970) pp. 589
164. Y. H. Ohtsuki, K. Koyama, Y. Yamamura: Phys. Rev. B **20**, 5044 (1979)
165. H. Winter: J. Phys.: Condens. Matter **8**, 10149 (1996)
166. R. Pfandzelter, J. Landskron: Phys. Rev. Lett. **70**, 1279 (1993)
167. K. Kimura, S. Ooki, G. Andou, K. Nakajima, M. Mannami: Phys. Rev. B **58**, 1282 (1998)
168. C. Lemell, J. Stöckl, J. Burgdörfer, G. Betz, HP. Winter, F. Aumayr: Phys. Rev. Lett. **81**, 1965 (1998)
169. K. Kimura, S. Ooki, G. Andou, K. Nakajima: Phys. Rev. A **61**, 012901 (2000)
170. G. Andou, K. Nakajima, K. Kimura: Nucl. Instrum. Methods B **160**, 16 (2000)
171. J. W. Rabalais, H. Bu, C. D. Roux: Phys. Rev. Lett. **69**, 1391 (1992)

Index

Springer Tracts in Modern Physics

155 **High-Temperature-Superconductor Thin Films at Microwave Frequencies**
By M. Hein 1999. 134 figs. XIV, 395 pages

156 **Growth Processes and Surface Phase Equilibria in Molecular Beam Epitaxy**
By N.N. Ledentsov 1999. 17 figs. VIII, 84 pages

157 **Deposition of Diamond-Like Superhard Materials**
By W. Kulisch 1999. 60 figs. X, 191 pages

158 **Nonlinear Optics of Random Media**
Fractal Composites and Metal-Dielectric Films
By V.M. Shalaev 2000. 51 figs. XII, 158 pages

159 **Magnetic Dichroism in Core-Level Photoemission**
By K. Starke 2000. 64 figs. X, 136 pages

160 **Physics with Tau Leptons**
By A. Stahl 2000. 236 figs. VIII, 315 pages

161 **Semiclassical Theory of Mesoscopic Quantum Systems**
By K. Richter 2000. 50 figs. IX, 221 pages

162 **Electroweak Precision Tests at LEP**
By W. Hollik and G. Duckeck 2000. 60 figs. VIII, 161 pages

163 **Symmetries in Intermediate and High Energy Physics**
Ed. by A. Faessler, T.S. Kosmas, and G.K. Leontaris 2000. 96 figs. XVI, 316 pages

164 **Pattern Formation in Granular Materials**
By G.H. Ristow 2000. 83 figs. XIII, 161 pages

165 **Path Integral Quantization and Stochastic Quantization**
By M. Masujima 2000. 0 figs. XII, 282 pages

166 **Probing the Quantum Vacuum**
Pertubative Effective Action Approach in Quantum Electrodynamics and its Application
By W. Dittrich and H. Gies 2000. 16 figs. XI, 241 pages

167 **Photoelectric Properties and Applications of Low-Mobility Semiconductors**
By R. Könenkamp 2000. 57 figs. VIII, 100 pages

168 **Deep Inelastic Positron-Proton Scattering in the High-Momentum-Transfer Regime of HERA**
By U.F. Katz 2000. 96 figs. VIII, 237 pages

169 **Semiconductor Cavity Quantum Electrodynamics**
By Y. Yamamoto, T. Tassone, H. Cao 2000. 67 figs. VIII, 154 pages

170 **d–d Excitations in Transition-Metal Oxides**
A Spin-Polarized Electron Energy-Loss Spectroscopy (SPEELS) Study
By B. Fromme 2001. 53 figs. XII, 143 pages

171 **High-T_c Superconductors for Magnet and Energy Technology**
By B. R. Lehndorff 2001. 139 figs. XII, 209 pages

172 **Dissipative Quantum Chaos and Decoherence**
By D. Braun 2001. 22 figs. XI, 132 pages

173 **Quantum Information**
An Introduction to Basic Theoretical Concepts and Experiments
By G. Alber, T. Beth, M. Horodecki, P. Horodecki, R. Horodecki, M. Rötteler, H. Weinfurter, R. Werner, and A. Zeilinger 2001. 60 figs. XI, 216 pages

174 **Superconductor/Semiconductor Junctions**
By Thomas Schäpers 2001. 91 figs. IX, 145 pages

175 **Ion-Induced Electron Emission from Crystalline Solids**
By Hiroshi Kudo 2002. 85 figs. IX, 161 pages

MIX
Papier aus verantwortungsvollen Quellen
Paper from responsible sources
FSC® C105338

If you have any concerns about our products,
you can contact us on
ProductSafety@springernature.com

In case Publisher is established outside the EU,
the EU authorized representative is:
Springer Nature Customer Service Center GmbH
Europaplatz 3, 69115 Heidelberg, Germany

Printed by Libri Plureos GmbH
in Hamburg, Germany